INNOVATIONS IN GIS 8

Spatial Information and the Environment

INNOVATIONS IN GIS 8

Spatial Information and the Environment

Edited by
Peter J Halls
Computing Service
University of York

London and New York

First published 2001 by Taylor & Francis
11 New Fetter Lane, London EC4P 4EE

Simultaneously published in the USA and Canada
by Taylor & Francis Inc,
29 West 35th Street, New York, NY 10001

Taylor & Francis is an imprint of the Taylor & Francis Group

© 2001 Taylor and Francis

Every effort has been made to ensure that the advice and information in this book is true and accurate at the time of going to press. However, neither the publisher nor the authors can accept any legal responsibility or liability for any errors or omissions that may be made. In the case of drug administration, any medical procedure or the use of technical equipment mentioned within this book, you are strongly advised to consult the manufacturer's guidelines.

British Library Cataloguing in Publication Data
A catalogue record for this book is available from the British Library

Library of Congress Cataloging in Publication Data

ISBN 0-415-25362-4

Contents

Preface

As we enter the new millennium, continued buoyancy in innovation in Geographic Information Science (GIS) is demonstrated by the presentations at the GIS Research UK 2000 (GISRUK 2000) conference which reflected a wide range of interests in GIS from numerous disciplines. The conference attracted far more submissions of a very high standard than could possibly fit into the two day format. From this wealth have been selected for this volume a thematic set of papers which are concerned with innovation in GIS in the study of the environment. These papers form a series from issues discussing the acquisition of information about the environment, exploration and analysis of that information and the application of the conclusions reached in decision support. In considering the role of environmental spatial information in decision support, this set expands upon the standard reference for environmental GIS, Goodchild, Steyaert and Parks (1996).

The scene is set by the opening chapter of this volume, Craig Trotter et al's discussion and challenge regarding spatial information for environmental modelling. This is a formalised version of Craig's guest presentation which opened the conference in style. In the environmental disciplines especially, many workers have to design and implement the entire range of a project, from data selection and acquisition through modelling and analysis to inform policy and management. Whilst offering some elements of control over the project such entirety makes demands upon the investigator and methodologies that comprise challenges for GIS. The remaining chapters explore various aspects of these processes and challenges, offering some innovative solutions and defining additional research agenda for the GIS community to tackle.

It has become traditional for a prize to be awarded for the best paper presented by a young researcher. The task of the review jury was especially hard this year, with several outstanding contenders. The eventual winner was Hartwig Hochmair and his paper is included in this volume. Conference participants were also able to vote for the best paper overall - this being won by Mathieu Barrault, whose paper is included with other papers from GISRUK 2000, primarily forming a Geo-computational set, in a companion issue of Computers, Environment and Urban Systems.

GISRUK 2000 was the eighth in the series and, in terms of both the material submitted and those who participated, continued the trend towards being a truly European event with contributions from throughout the world. This balance is reflected in this volume, with over a third of the contributions being from outside the UK. The cross-fertilisation of ideas that results from such a multi-national gathering bodes extremely well for future research and collaboration in the GIS community as a whole.

Once again, postgraduate students commencing their research careers in GIS were able to meet together with experienced workers and publication editors prior to the formal conference. This event is designed to introduce the participants to issues relating to the UK research environment, to the requirements that must be met in order to get their work published, to foster the principles of Continuing Professional Development and to enable research students from different institutions to compare notes. It is hoped that the Young Researcher's Forum will now become a standard part of the GISRUK event.

The quality of submissions continues to rise: many more papers were deemed by the peer review process to be acceptable than the conference format could support; all papers were reviewed by three independent reviewers on submission. Final acceptance was in part governed by the number of thematic sessions that could fit in the conference format, and a significant number of otherwise excellent submissions had to be declined because there were too few other related papers to make a viable session. Those papers included in this volume, and its companion in the CEUS series, have been recommended and further reviewed by appropriate independent experts.

The GISRUK series is dependent upon many people, sponsors, reviewers, speakers, stewards, and so on. Whilst it is impossible, and inappropriate in this volume, to name them all there are a number whom I must single out for explicit thanks. The (UK) Association for Geographic Information (AGI) sponsored Professor Waldo Tobler (UCSB) to come to York as one or our guest speakers and also enabled students to participate at a reduced rate. Taylor & Francis and Whitbread Plc enabled Craig Trotter to come over from New Zealand as our other guest speaker. Know Edge Ltd funded the Young Researcher's Prize and Autodesk (UK) provided the prize for the best paper overall. The AGI and RRL.net enabled the Young Researcher's Forum and the partnership of CEUS and, last but by no means least, Wensleydale Dairy Products Ltd provided a social event that demonstrably proved to be a Grand Evening In!

There are individuals who must be mentioned, the national GISRUK Steering Committee and the local University of York Organising Committee members have been extremely supportive throughout the period involved. Lorraine Moor designed and set out the Conference Proceedings; Mary Kennedy, Jenny Hammond and Louise Scholes have spent long hours in getting this volume together. Paul Longley, Editor in Chief of CEUS, and Tony Moore, of Taylor and Francis, have been instrumental in bringing about the conference edition of CEUS and this volume respectively. The University of York, and particularly the Archaeology, Biology, Computer Science and Environment Departments and my own Computing Service colleagues have contributed a great deal.

To all my grateful thanks.

In 2001 GISRUK makes its first visit to Wales, to be hosted jointly by the universities of Glamorgan and Cardiff, and in 2002 it is planned to return to Yorkshire, to Sheffield. I look forward to meeting you there.

Peter Halls
York September 2000

Sponsors

AGI
Blackwells University Bookshop (York)
ESRI (UK)
Know Edge Ltd.
Stockholm Environmental Institute at York
Whitbread Plc

Blackwells Publishers
CEUS
John Wiley Publishers
RRL.net
Taylor & Francis Ltd.
Wensleydale Dairy Products Ltd.

GISRUK Committees

GIS Research UK Steering Committee

Steve Carver	University of Leeds
Jane Drummond	University of Glasgow
Bruce Gittings (Chair)	University of Edinburgh
Peter Halls	University of York
Gary Higgs	University of Wales at Cardiff
Zarine Kemp	University of Kent
David Kidner	University of Glamorgan
David Martin	University of Southampton
George Taylor	University of Newcastle upon Tyne
Steve Wise	University of Sheffield

York Organising Committee

Steve Cinderby	Steve Dobson
Rob Fletcher	Peter Halls (Chair)
Colin McClean	Lorraine Moor
Fiona Polack	Julian Richards
Piran White	

Contributors

Frederic Barbe
French Naval Academy, SIGMA Laboratory, Lanveoc Poulmic, BP 600 29 240 Brest Naval, France.

Jaishree Beedasy
Faculty of Engineering University of Mauritius
(jaish@uom.ac.mu)

Rebekah Boott
Centre for Advanced Spatial Analysis, University College London, 1-19 Torrington Place, Gower Street, London WC1E 6BT
(hannah.boot@ucl.ac.uk)

Christophe Claramunt
Nottingham Trent University, Department of Computing, Burton Street, Nottingham NG1 4BU
(clac@doc.ntu.ac.uk)

A.J.Comber
Macaulay Land Use Research Institute, Land Use Science Group, Craigiebuckler, Aberdeen, AB15 8QH
(l.comber@mluri.sari.ac.uk)

Morten Dæhlen
Department of Informatics, University of Oslo, P.O. Box 1080, Blindern, 0316, Oslo, Norway
(mod@forskningsradet.no)

Thomas Devogele
French Naval Academy, SIGMA Laboratory, Lanveoc Poulmic, BP 600 29 240 Brest Naval, France.
(devogele@ecole-navale.fr)

Morten Fimland
Department of Informatics, University of Oslo, P.O. Box 1080, Blindern, 0316, Oslo, Norway

Robert Frank
Computing Laboratory, University of Kent, Canterbury, Kent CT2 7NF
(rf7@ukc.ac.uk)

François Gélébart
French Naval Academy, SIGMA Laboratory, Lanveoc Poulmic, BP 600 29 240 Brest Naval, France.

Verena Gruener
Frauenlobstr. 22, 80337 Muenchen, Germany
(verena_gruener@hotmail.com)

Mordechai Haklay
Centre for Advanced Spatial Analysis, University College London, 1-19 Torrington Place, Gower Street, London WC1E 6BT

Peter J Halls
Computing Service, University of York, Heslington, York, YO10 5DD
(P.Halls@york.ac.uk)

Richard J. Harris
School of Geographical Sciences, University of Bristol, University Road, Bristol BS8 1SS (Now at Centre for Advanced Spatial Analysis, University College London, 1-19 Torrington Place, Gower Street, London WC1E 6BT)
(R.J.Harris@bris.ac.uk)

Kate Heppell
Centre for Advanced Spatial Analysis, University College London, 1-19 Torrington Place, Gower Street, London WC1E 6BT

Marjan van Herwijnen
Institute for Environmental Studies, Vrije Universiteit Amsterdam, De Boelelaan 1115, 1081 HV Amsterdam, The Netherlands
(marjan@ivm.vu.nl)

Øyvind Hjelle
Department of Informatics, University of Oslo, P.O. Box 1080, Blindern, 0316, Oslo, Norway

Hartwig Hochmair
Department of Geoinformation, Technical University of Vienna, Gusshausstr. 27-29, A-1040 Wien, Austria
(hochmair@geoinfo.tuwien.ac.at)

Howard, D.C
Centre for Ecology & Hydrology, Merlewood, Grange-over-Sands, Cumbria, LA11 6JU

Jasmee Jaafar
School of Geography, The University of Nottingham, Nottingham, NG7 2RD

Ron Janssen
Institute for Environmental Studies, Vrije Universiteit Amsterdam, De Boelelaan 1115, 1081 HV Amsterdam, The Netherlands

Sven D. Jelaska
Department of Botany, Faculty of Science, University of Zagreb, Marulicev trg 20/2, HR-10000, Zagreb, Croatia
(sven@croatica.botanic.hr)

Taskin Kavzoglu
School of Geography, The University of Nottingham, Nottingham, NG7 2RD
(kavzoglu@geography.nottingham.ac.uk)

Zarine Kemp
Computing Laboratory, University of Kent, Canterbury, Kent CT2 7NF

David M. Kidd
Department of Geography, University of Portsmouth, Buckingham Building, Lion Terrace, Portsmouth PO1 3HE
(david.kidd@port.ac.uk)

A. N. R. Law
Macaulay Land Use Research Institute, Land Use Science Group, Craigiebuckler, Aberdeen, AB15 8QH

J.R. Leathwick
Landcare Research, Private Bag 11 052, Palmerston North, New Zealand

J. R. Lishman
Department of Computing Science, University of Aberdeen, AB24 3UE

Paul A. Longley
School of Geographical Sciences, University of Bristol, University Road, Bristol BS8 1SS (Now at Centre for Advanced Spatial Analysis, University College London, 1-19 Torrington Place, Gower Street, London WC1E 6BT)

Paul M. Mather
School of Geography, The University of Nottingham, Nottingham, NG7 2RD

David Miller
Macaulay Land Use Research Institute, Craigiebuckler, Aberdeen, AB15 8QH
(d.miller@mluri.sari.ac.uk)

Jeremy Morley
Centre for Advanced Spatial Analysis, University College London, 1-19 Torrington Place, Gower Street, London WC1E 6BT

D. Pairman
Landcare Research, Private Bag 11 052, Palmerston North, New Zealand

Rameshsharma Ramloll
Computing Science Department, University of Glasgow, G12 8QQ

Scott, W.A
Centre for Ecology & Hydrology, Merlewood, Grange-over-Sands, Cumbria, LA11 6JU

C.M. Trotter
Landcare Research, Private Bag 11 052, Palmerston North, New Zealand
(trotterc@landcare.cri.nz)

C.A.O. Vieira
School of Geography, The University of Nottingham, Nottingham, NG7 2RD
(vieira@geography.nottingham.ac.uk)

Watkins, J.W
Centre for Ecology & Hydrology, Merlewood, Grange-over-Sands, Cumbria, LA11 6JU
(jww@ceh.ac.uk)

Introduction

Section 1 Information Acquisition in the Environment

"Environmental models require data for calibration, verification, and specification of boundary conditions. The lack of suitable data has been one of the most serious impediments to the development of environmental models." (Goodchild, Steyaert, Parks, et al, 1996 p1)

As Trotter et al state in the abstract of their keynote paper, *Environmental policy and management is increasingly required to optimise both the use of ecosystems for commodity supply and the maintenance of ecosystem function and viability to ensure continuity of life supporting services. This optimisation can not be achieved without access to improved information on ecosystem state and performance at landscape scales.* So has nothing changed since the Second International Conference / Workshop on Integrating Geographic Information Systems and Environmental Modelling, in 1993. Trotter et al demonstrate that there has, indeed, been a great deal of progress over those seven years but also identify challenges that need addressing in the future. Comber et al explore methodologies for the automation of Land Cover change detection, whilst Harris and Longley explore Urban Environments and challenges in drawing together environmental information with the wealth of other data describing urban characteristics.

Field boundaries are vital landscape features for many species, in addition to being a factor that makes a landscape attractive. Kavzoglu, et al, explore methodologies for the extraction of field boundary information from classified satellite imagery.

We place a great deal in the reliability of our classified satellite imagery but, to date, have had few techniques for assessing their spatial reliability rather than the accuracy of classification. Vieira and Mather conclude this section with a methodology for assessing the spatial reliability of thematic images.

Section 2 Manipulating the information: Tools, Visualisation and Navigation

"The availability [in GIS] of different depictions and different perspective views of the same dataset is assumed to be essential. ... The design of visualizations is a critical aspect of GIS, especially where these images are to be employed as spatial decision support systems (SDSS)." (Wood and Brodlie, 1994, p7)

The tools to manage, manipulate and visualise the information maintained within a GIS are crucial to the support of analysis. As diverse disciplines seek to understand their own studies better through the application of techniques of spatial analysis so the demands for novel methods to support this analysis increase.

Daehlen, et al, seek to address the problems of the representation and maintenance of multi-resolution information. They propose a triangle-based representation of surfaces together with the integration of curve networks as a means of managing the interrelationships of the surface to linear features associated with it.

When studying environmental processes it is essential that the methodologies employed take into account time as well as space. Techniques for the manipulation and analysis of spatialtemporal information are the focus of ongoing work by Frank and Kemp, who propose a flexible component based framework to support environmental study.

Navigation control systems, especially where the purpose is to detect and avoidpotential collisions, are another example of spatiotemporal analysis but with a time-critical component. Barbe, et al, present work on the development of a marine navigation / collision detection – avoidance system from the French National Naval Academy.

Hartwig Hochmair, the winner of the Best Paper by a Young Researcher award, asks us to adopt an alternative cognitive approach, arguing that in order to model movement in a representation of an environment it is also necessary to model the accuracy of that representation. He explores the differences between the real world and its representation and the effects these differences have on navigation strategies.

Section 3 Analysing the information: Computation and Modelling

Most definitions of GIS include concepts of modelling spatial activity and the analysis of the interactions of the represented real world objects over space (and sometimes time). Burrough and McDonnell (1998) cite Parker's 1988 definition of a GIS as "*an information technology which stores, analyses, and displays both spatial and non-spatial data*". Modelling is fundamental to environmental analysis: seventy percent of Goodchild et al (1996) is devoted to environmental modelling and GIS. In this section, Gruener adopts the established cellular automata methodology to build a dynamic model to study plant dispersal over space and time whilst Jelaska reports the beginnings of work to model and study patterns of forest plant diversity in the Nature Park on Mt Medvednica, near Zagreb, Croatia.

Environmental modelling has frequently to interpolate information between observations and is concerned with a multiplicity of factors. Kidd, for his study of the Saddle-backed bushcricket, reviews commonly employed interpolation techniques and then demonstrates his own, hybrid, proposal.

This section is concludes with Miller's work on the modelling of the visibility of land use in an environmentally sensitive area of Scotland. This work is intended to assist in the understanding of the way land use contributes to and controls the landscape and to assist planners in determining appropriate development policies.

Section 4 Applying the information: Decision Support.

Environmental data is collected and environmental spatial modelling undertaken for a purpose. Sometimes that purpose is to understand the ecology of some species or community, other times it is to inform decision making processes that may threaten an environment or some part thereof. Cowen concluded in 1988 that GIS is: "*a decision support system involving the integration of spatially referenced data in a problem solving environment*". (Cowen 1988)   Trotter, et al, in the opening chapter of this book, are concerned with ecosystem fucntion and viability to ensure the continuity of life supporting services. Their work is concerned with the whole cycle of environmental spatial modelling, data collection, manipulation, analysis and forecasting: enabling the decision making process to maintain the viability of the ecosystem, the environment in which we live. In the preceding chapter, Miller has discussed methodologies for the modelling and visualisation oflandscape factors to support decision making. This section of four chapters is concerned specifically with spatial decision support.

Beedasy and Ramloll are concerned with the mechanisms by which users interact with GIS, or spatial decision support system (SDSS), in order to promote the participation of the members of a local community in the decision making process. They explore the application of research in Computer Supported Cooperative Work (CSCW) to spatial decision making processes.

Boott, et al, consider the role GIS can play in enabling the reuse of industrial land, *brownfield* redevelopment. In this work they demonstrate the value of Exploratory Spatial Analysis (ESA) to support public participation in the decision making process.

van Herwijnen and Janssen return to the consideration of the users of SDSS in their discussion of evaluation methods in spatial multi-criteria analysis and propose a methodology based on *performance maps*.

Finally, Howard, et al, discuss the use of statistical classification in the determination of (spatial) change in the components that make up an ecosystem. The methodology they describe provides for the calibration of their classification metrics in order to enable the assessment of the likelihood of change.

This progression of information acquisition, manipulation and analysis leads to one of the ultimate purposes of undertaking environmental analysis: the informing of the decision making process such that the impact of (human) intervention may be measured, perhaps mitigated, and the environment, upon which we depend, conserved. This small volume makes an up to date contribution to the progress of the application of GIS in environmental analysis and also sets out new challenges for the research agenda of the future.

References

Burrough, P.A., & McDonnell, R.A., 1998, *Principles of Geographical Information Systems.* Oxford University Press, Oxford.

Cowen, D.J., 1988, GIS versus CAD versus DBMS: what are the differences? *Photogrammetric Engineering & Remote Sensing*, 54, pp 1551-1555.

Goodchild, M F, Steyaert, L T, Parks, B O, Johnston, C, Maidment, D, Crane, M, and Glendinning, S (Eds) 1996, *GIS and Environmental Modeling: Progress and Research Issues*. GIS World Books, Inc, Fort Collins.

Wood, M., and Brodlie, K., 1994, ViSC and GIS: Some fundamental considerations. *In* Hearnshaw, H.M. & Unwin, D.J.,, 1994, *Visualization in Geographical Information - Systems*. John Wiley & Sons, Chichester.

Section 1

Information Acquisition in the Environment

1

Spatial Information For Ecosystem Classification, Analysis, And Forecasting

C.M. Trotter, J.R. Leathwick, and D. Pairman

Environmental policy and management is increasingly required to optimise both the use of ecosystems for commodity supply and the maintenance of ecosystem function and viability to ensure continuity of life supporting services. This optimisation can not be achieved without access to improved information on ecosystem state and performance at landscape scales. We consider here approaches to delivering enhanced information by combining remote sensing with ecological modelling and prediction. The advantages and limitations of statistical, process-based, and knowledge-based models are discussed, using examples from two important issues of international environmental concern: forest biodiversity, and carbon sequestration. Current limitations on the availability and accuracy of spatial information to address these and similar issues are identified. Possibilities for removing these limitations through advances in optical and radar remote-sensing technology are presented, and also reviewed are recent advances in the processing of remotely sensed data to deliver more accurate biophysical information. We conclude that although much progress can yet be made in ecosystem classification, modelling, and forecasting, more attention must be given to removing the shortcomings in the scale and accuracy of spatial data, and also in the specification of spatial error-particularly for the climate and soils spatial datasets that underpin ecological analysis at landscape scales. Also required is more investigation of the advantages and limitations of the various analysis and modelling approaches applied to ecological datasets, given the apparent divergence in results when differing analysis methods are applied to the same dataset.

1. Introduction

In the last decade it has become widely acknowledged that human activity has the potential to change significantly, and perhaps irretrievably, the nature of the environment even at global scales (IPCC 1995, Gower *et al.* 1999). Resource depletion and degradation are becoming sufficiently widespread that serious questions are being raised about the degree to which we are living at the expense of future generations. Despite the undeniable pressures on the environment exerted by human activities, only limited success has been achieved to date in developing policy frameworks that limit damage to, and fully value, the ecosystems mankind ultimately depends on for life-supporting capacity (Daly 1997).

It is clear that in the future society will have to give more urgent attention to the conflicts inherent in using ecosystems for resource and commodity supply, while also needing to maintain these same ecosystems as vital yet finite suppliers of life-supporting services. To achieve this, environmental policy, management, and information will have increasingly to be directed towards delivering more carefully optimised solutions that protect ecosystems while ensuring socio-economic development. Policy development will need to adopt a more precautionary approach, as knowledge of ecosystem function, response, and viability is presently limited, and the opportunities for recovering from poor decisions are becoming few. Both policy and management will also need to take account of quantitative environmental information and research to underpin decisions. Research, too, will have to become more responsive, to address in a more timely manner the information needs of environmental policy and management (Walker and Young 1997). These trends in environmental policy and management are already being seen in the international arena, in the collaborative political, policy, and scientific effort surrounding the development of the Kyoto Protocol (IGBP 1998): a precautionary approach to the potential problem of greenhouse gas-induced climate change.

Improved environmental management requires action at landscape scales (Walker and Young 1997). To support this, better information will be required in both space and time, with proper acknowledgement of the intricacies of ecosystem interactions and interdependencies (Karveia 1994, Farina 1998, Waring and Running 1998). However, in this we face a considerable challenge. Most current knowledge of ecosystem function and viability has been developed from plot-scale ecosystem research. Efforts to develop quantitative approaches for spatial generalisation of this plot-scale knowledge have gained momentum only relatively recently (e.g. Austin and Cunningham 1981, Scepan *et al.* 1987, Yee and Mitchell 1991, Aspinall and Veitch 1993, Leathwick *et al.* 1996, Waring and Running 1998, Hall and Hollinger 2000). At least part of the reason for this has been a perceived lack of urgency: it is only since 1970 that concern has been formally raised that ecosystems may in the future fail in the delivery of a range of life-support services to humans (Ehrlich and Ehrlich 1970, SCEP 1970). Another significant factor is that the spatial information technologies necessary to analyse and model the complexities of ecosystems in space and time have only become widely available since the 1980s. It is not surprising, then, that our understanding remains limited of the way in which natural ecosystems function, respond and retain viability at landscape scales.

In this paper we review progress made, and issues yet to be addressed, in developing spatial information to answer questions related to the state, function, response, and viability of terrestrial ecosystems at landscape scales. We begin by examining the evolving role of spatial information for ecosystem description and analysis, and assess information availability and shortfalls against the requirements for improved environmental management. The current use and limitations of spatial models for analysing and forecasting of ecosystem state and function are then considered, taking examples from two topical areas of international environmental concern: the assessment of terrestrial vegetation biodiversity, and carbon sequestration in forests. Finally, the role that continuing development in remote sensing will play in our ability to monitor ecosystem state over wide areas is discussed. This includes an overview of current advances in sensor technology and data processing for improved measurement of vegetation biophysical parameters from space.

2 The Evolving Role Of Spatial Information In Environmental Analysis

2.1. Information Requirements and Current Data

Optimised environmental decision-making requires, at the very least, access to databases that are spatially comprehensive, current, relevant, and thematically detailed enough to allow the distribution and state of ecosystems to be defined. Newly acquired data should be backwards-compatible with existing data, to allow detection of ecologically important change. Information describing state and change should be linked to processes that govern ecosystem physical function and viability, and quality as a life-support system. The process-based link is of particular importance when attempting to forecast the probable impact of pressures on the environment, or of management intervention, under non-equilibrium conditions. Such conditions already exist for the vegetative component of many ecosystems today, due both to direct physical disturbance by man and to the indirect effects that increased concentrations of atmospheric nitrogen and sulphur compounds, and CO_2 levels, have on the nutrient and carbon cycles (e.g. Gower *et al.* 1999)

Unfortunately, the foregoing statements about an ideal set of terrestrial environmental information contrast rather strongly with current reality. Present terrestrial ecosystem datasets are often fragmentary, are frequently dispersed across a number of agencies, and data are seldom collected uniformly in either geographic or environmental space. The datasets in many instances relate largely to resource inventory, and rest principally on a history of documenting resource availability for use. Such information is well suited to answering questions relating to location, area, shape, and spatial context. However, it is generally much less suited to determining ecosystem characteristics now becoming more important, those of functional performance, viability, and resilience to disruption-information on current vegetation composition and extent being probably the only notable exception. Neither are the requirements for more accurate forecasting of ecosystem trends easily met by information in existing spatial datasets. Forecasting ideally requires process-linked information on the water, nutrient, and carbon cycles that underpin ecosystem function and viability. Although many datasets describing natural resources do contain data on soil, climate, and plant biomass parameters, these are generally in the form of sparse point datasets. Forming continuous surfaces from these for spatial prediction is a process whose accuracy remains something of an open question, especially in areas with complex topography or geomorphology.

Although it is easy to be critical of the state of our ecological and physical resource information at landscape scales, there are limited possibilities for obtaining relevant spatially continuous data by traditional methods based on ground survey and image interpretation. Much of the information required to address current and future environmental issues can not be directly inferred from imagery, particularly information related to the water, nutrient, and carbon cycles, or to the fluxes in these quantities that can provide snapshots of ecosystem performance. Furthermore, simple logistics will defeat the use of more intensive ground-based sampling schemes to generate continuous surfaces of thematic data for such variables, particularly in steeplands where much of the world's biodiversity resides and strong climatic gradients create complex soil and vegetation patterns. Although quite coarse sampling can deliver reliable average values for parameters, this will not usually be sufficient for environmental monitoring. Often it will be localised change in zones of reduced ecosystem viability, change that has little influence on the mean value for a wide area, that will provide the most timely signal of impending ecological failure. Obtaining meaningful environmental information over wide areas, while retaining sufficient detail to represent properly and track local ecosystem complexity and change, is therefore largely an impossible task if considered in terms of well-established resource mapping or sampling methods.

2.2. Alternative Approaches to Ecosystem Depiction and the Role of Models

Given the problems outlined above, we clearly require new approaches to the classification and analysis of ecosystem resources, and evaluation of their functional state. The key to success in landscape-scale monitoring and evaluation is deceptively simple: we need only link ecosystem state, function and viability to observable parameters. Unfortunately, the critical parameters that have been developed to optimally monitor ecosystem performance at plot scales are often not directly observable at landscape scales (e.g. Farina 1998, Waring and Running 1998). So alternative approaches to ecosystem characterisation must be devised. Remote sensing has a special role to play in this. It is the only approach that is capable of providing information on ecosystem state over large areas at reasonable cost, while retaining an appreciable amount of the detail of landscape complexity. However, it is also well known that there are many parameters that have traditionally been measured in plot-scale ecosystem studies that will never be observable by remote sensing.

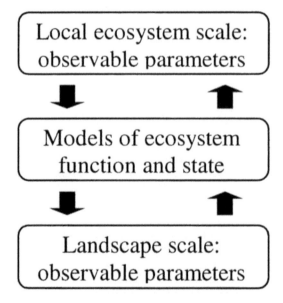

Figure 1. Ecological models: the link between plot-scale data on ecosystem state and function, and landscape-scale parameters observable using remote sensing.

To overcome this, remote sensing must in the future be oriented much more towards providing robust measures of parameters that can be compared with predictions of ecosystem state and performance made using proven ecosystem and ecophysiological models (Figure 1). Alternatively, remote-sensing can be used to provide spatially explicit input data for models proven at plot scales, provided these models are formulated in such a way as to accept such parameters as can be readily observed (Figure 1). In particular, remote sensing will need to produce improved data to construct and/or validate climate surfaces, to measure vegetation light absorption and biomass parameters, and to determine plant photosynthetic activity. Photosynthetic activity, an important component of ecosystem viability in its own right, can also potentially be linked to the availability of soil water and nutrients, and so help to validate spatial estimates of these otherwise difficult to establish parameters. Improved discrimination of vegetation composition will also continue to be a very important goal for remote sensing development. However, emphasis should be placed more on robust discrimination of those groups of species that fulfil differing functional roles in terms of ecosystem viability and performance, with less effort on trying to meet the generally greater levels of individual species discrimination formerly achieved by mapping based on the combination of field data and stereo photo interpretation.

In a future of more considered environmental management, ecosystem models will play a much larger role than simply providing the link between plot-scale ecosystem knowledge and landscape-level environmental information and monitoring. Optimised environmental management will demand a highly multidisciplinary approach to environmental problem solving, and ecosystem simulation models will make an essential contribution to this. Such models provide a formal framework within which to assemble and test data, theories, and opinions, over a range of scales in space, time, and biological complexity. Process-based models provide probably the only means to estimate ecosystem functional state forwards and backwards in time under non-equilibrium conditions. These models have the advantage that they can be evaluated for reliability by using historical data to predict current status and trend, which can then be validated against current observations. Any quantitative model also provides the usual advantages of a formal means of assessing and optimising alternative intervention strategies, and determining sensitivity to the reliability of input data-characteristics that are especially valuable when making environmental decisions on the basis of limited data.

3. Progress In Ecosystem Classification And Analysis At Landscape Scales

In general, there is no single approach to ecological analysis at landscape scales. Different questions will require that different variables be estimated, to differing degrees of accuracy, and with differing extents of spatial and temporal interconnectedness (Levin 1992, Farina 1998, Waring and Running 1998). However, the common theme running through these analyses is that ecosystems are very complex systems for which limited information presently exists. This requires, whenever possible, that ecosystem analysis be simplified through a careful understanding of ecological principles and processes. The three most common approaches to providing landscape-scale information on ecosystems from analysis of plot-scale ecological data involve the use of statistical, process-based, and knowledge-based models. It is not our intention to provide a comprehensive overview of these methods, but instead to use selected examples to outline the basic assumptions and simplifications in these different modelling approaches, and to summarise both current progress and challenges yet to be resolved.

3.1. Statistical Models

Statistical models have become used increasingly for predicting the broad-scale distribution of plant species (e.g. Austin and Cunningham 1981, Austin *et al.* 1990, Leathwick 1995), and also of animals (e.g. Pereira and Itami, 1991). For a given species, plot-scale training data, describing either species presence/absence, or abundance, are used to determine the primary environmental correlates that influence distribution. In the case of animal populations, additional variables such as the quality of available habitat for forage and shelter will usually be required to represent the distribution adequately. The analytical methods employed to describe species distributions in multi-variate environmental/habitat space range from simple regression, to the generalised non-parametric fitting technique of generalised additive models, or GAMs (e.g. Hastie and Tibshirani 1990, Yee and Mitchell 1991, Leathwick 1995). Non-parametric techniques such as GAMs are preferred for ecosystem analysis, as plant and animal populations seldom exhibit a parametric response to environmental or other variables.

Regardless of the analysis method used, scaling from plot-based observations to the landscape-level requires spatial layers of the predictor variables. In the case of plant species distributions, considered for the remainder of this section, these are typically some combination of such physiologically important parameters as solar energy, temperature, soil water availability, and soil fertility. Figure 2 shows an example of using such variables to predict the potential indigenous forest pattern for New Zealand, at a 1 km grid scale. The predicted tree distributions reflect not just the effects of environmental drivers, but also competitive interactions between the major species (Leathwick 1998, Leathwick and Austin 2000). The data were created from GAM analyses of compositional data from 14,500 plots describing the abundance of major tree species. Spatial layers of the climate variables for the prediction were established by fitting thin-plate splines to irregularly distributed climate station data (Hutchinson and Gessler 1994, Leathwick and Stephens 1998). Soil water and nutrient spatial data were derived from a 1:1,000,000 scale soil map, together with a spatial water balance model (Leathwick *et al.* 1998). As with many predictions of this type, the accuracy of the final product is difficult to assess rigorously, because of the largely unknown limitations in the spatial data layers used for prediction. However, error estimates derived from the regression models suggest that standard errors for most species range within 10–20% of their predicted mean abundance.

Basing environmental policy and management action on such data as shown in Figure 2 is perhaps less than ideal, but is a much better alternative than so often prevails: proceeding with little or no spatially comprehensive data to inform decisions being taken. Such data are already being used, for example, to prioritise the acquisition and management of conservation lands so that viable ecological reserves are well distributed over environmental space (Margules and Pressey 2000)- rather than simply spread over geographic space as has often been the approach in the past. When the predictions of vegetation composition are derived from plot data collected at reasonably undisturbed sites, information on potential vegetation composition can form a baseline for the assessment of ecosystem change. For example, deviations from the dominant species predicted for an area can be used as a measure of change in ecological state, and may also have implications for a change in ecosystem function, from some desirable benchmark. Changes from conifer-dominated to a broadleaf dominated forest fall into this category, and have implications for such things as carbon sequestration, nutrient cycling, and

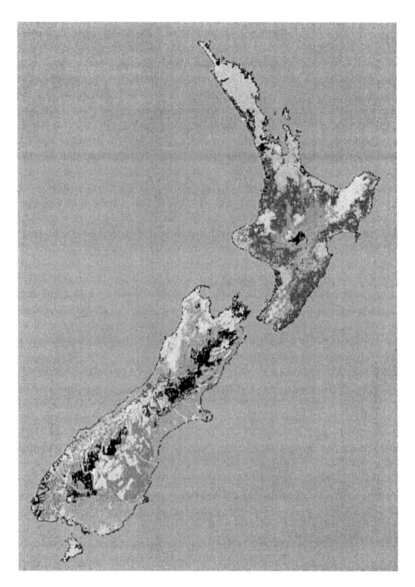

Figure 2. Potential forest cover predicted in 20 classes (each comprising a mixed species group) from a GAMS analysis of 14,500 species composition plots using climate and soils parameters as explanatory variables, and extrapolated to landscape scales using spatial climate and soil datasets. A detailed description of the classes and their composition is given in Leathwick (2000).

habitat quality. Species richness can also be modelled (Leathwick *et al.* 1998), with differences in richness between potential and actual measures clearly having implications for biodiversity management, and probably also for ecosystem function. Analysis can also be extended to incorporate the effects of large-scale historic disturbances on species distributions (Leathwick 1998), and of inter-specific competition (Leathwick and Austin 2000).

The ability to predict the distribution of vegetation at landscape scales fits well with the requirements for better utilising remote sensing for environmental monitoring. Information on broad canopy vegetation composition, readily extractable from remotely-sensed data, can be considerably augmented using modelled species association data. Both co-occurring but spectrally indistinct canopy species, and largely undetectable sub-canopy species, can be added to descriptions of vegetation composition, provided that something is known of the disturbance history of the areas being considered. Generalised ecosystem characteristics such as species dominance or richness are also often more quantifiable than detailed descriptions of composition, while still providing important information that can be related to ecosystem state, function, and viability. An example of this provided in Plate 1. The area shown is montane conifer/broadleaved forest, parts of which have been seriously degraded over a 40 year period due at least partially to preferential browsing of some conifer species by the brush-tailed possum (*Trichosaurus vulpecula*). The result is an increase in the broadleaved to conifer tree fraction, and in some areas conifers are now absent. Deviations in both species dominance, and richness, have occurred. In this case, these changes can be quantified using remote sensing to measure the conifer canopy fraction for a given area (Plate 1). In terms of monitoring environmental trend, a fractional cover map is in this case more use than a more traditional vegetation map that would probably consist of just two classes separated by a somewhat arbitrary boundary: broadleaved, and conifer/broadleaved, forest. Much ecological change could occur within the conifer/broadleaved class without being recorded as a shift in the class boundary, whereas such change would certainly be recorded by a fractional cover parameter.

Data from statistical analyses of vegetation surveys continue to find expanding uses. Examples include predicting where undesirable invasive species are likely to establish, and assessing the impacts on vegetation composition of potential shifts in climate patterns (Leathwick *et al.* 1998, Rutherford *et al.* 1999). A very useful attribute of statistical models is that no *a priori* knowledge of the form of the relationship between plant or animal distribution and explanatory variables need be available, an advantage when attempting to represent the complex response of ecosystems. In general, the statistical approach is very powerful when training datasets adequately span the entire environmental/habitat space, and when disturbance at sites used to generate training data is minimal or has reached a steady state-although some caution is necessary as the impacts of disturbance events may span hundreds of years in the case of some forest species (e.g. Leathwick 1998).

3.2. Process-based Models

For many environmental issues there are relatively few data to support an analysis of ecosystem state and function, or the data may not be well distributed over environmental space. Disturbance may also make it difficult to obtain data that represent an equilibrium state. If an understanding of the basic processes controlling ecosystem development and change can be developed at the plot scale, in terms of variables for which spatial datasets exist, ecosystem state and viability can then be modelled and predicted at landscape scales. Process-based modelling approaches require considerably greater insight into the causal relationships that direct ecosystem state and change than do statistical models, but compensate through reduced requirements for training data. Properly formulated, process-based models can also perform reliably under a wider range of conditions than the training data are available for. We consider here the characteristics of two spatially distributed models, one for predicting forest carbon production and the other for forecasting forest species distribution and succession.

3.2.1   Modelling carbon production

The amount of carbon, more properly termed net primary production (NPP), produced by plants underpins the viability of most terrestrial ecosystems. Evidence is mounting that primary production can be linked to issues of biodiversity, resistance to disease, and the risk of invasion by exotic species (Adams and Woodward 1989, Waring and Pitman 1985, Stohlgren *et al.* 1999). A common method of estimating NPP makes use of the well-known concept that over long periods of time NPP is directly proportional to the amount of photosynthetically active radiation absorbed by a plant canopy (Monteith 1977):

$$NPP = \varepsilon \, \phi \qquad\qquad (1)$$

The term ε is the canopy light use photosynthetic efficiency, in units of mass of carbon produced per unit of absorbed photosynthetically active radiation (APAR), ϕ.

Equation (1) has been widely used to provide site-based values of NPP for grassland and forest ecosystems, and is also attractive for regional- to global-scale estimates because spatial measures of APAR can be obtained from remote sensing (e.g. Prince and Goward 1995, Knyazikhin *et al.* 1998, Gobron *et al.* 1999a). However, when estimating NPP over wide areas, it is necessary to recognise that ε is not actually a constant, as photosynthetic rates are constrained by both environmental and plant physiological parameters (e.g. Landsberg and Gower 1997). An improved formulation of equation (1) that takes these constraints into account is given by (Landsburg and Gower 1997, Landsburg and Waring 1997):

$$NPP = \varepsilon \, \phi f_D f_\theta f_T f_A \qquad\qquad (2)$$

The functions f_i account, respectively, for constraints on photosynthesis imposed by closing of stomata during periods of low humidity, limited soil water availability, low temperatures, and reductions in water flow within trees as they age. With this formulation, ε is largely independent of environmental conditions, and is reasonably consistent within broad forest classes (Landsburg and Waring 1997).
    To allow this model to be applied at landscape scales, a number of simplifications must be made. First, the forest physiological responses embedded in the constraint functions are based on generalisations of well established ecophysiological principles. This allows single functions to be reliably applied to all species within a broad forest type, all conifers say, even though explicit data for a particular species may not be available (Landsberg and Waring 1997). Second, spatial climate layers provide the plot-to-regional scale translation for the solar radiation, rainfall, temperature, and humidity parameters required to formulate the constraint functions at landscape scales (Coops *et al.* 1998). Spatial data on soil water holding capacity and nutrient status, and a broad estimate of stand age, are also required. Third, APAR is determined from suitable remotely sensed data, which also provides the necessary information on the location and area of the broad forest groups being considered. Once the amount of carbon produced has been calculated, it is then split between roots, stems, and leaves. This process again

makes use of simplified versions of well established principles of carbon allocation that depend on measures of environmental harshness, together with the adoption of generalised allocation partitioning equations for broad forest types (Landsberg and Waring 1997).

The constrained light use efficiency model described by equation (2) is both relatively simple and undemanding of input data when compared to most alternative process-based models currently available (e.g. FOREST BGC, Running and Gower 1991; BIOME BGC, Running and Hunt 1993). Nonetheless, it has performed remarkably well in site-based trials using local climate and soils data, in conjunction with physiological data broadly appropriate for the forest type being modelled (e.g. Landsberg and Waring 1997, Law *et al.* 1999, Landsberg *et al.* 2000). Because of its success at the site level, its use at landscape scales should depend primarily on the accuracy of the necessary spatial climate, soils, and APAR data, and also on the requirement to have an estimate of forest age. Sensitivity analyses suggest that the particular values of the physiological parameters play a lesser role in determining overall model accuracy than the probable inaccuracies in the spatial data, with the exception of the parameter that describes the sensitivity of stomatal closure to humidity. The extent of variation of this latter parameter between various broad forest types has yet to be established (Landsberg 2000), although recent work suggests that a generalised formulation of this parameter may yet be possible (Sperry *et al.* 1999).

An example of spatial prediction of annual NPP is shown in Plate 2. Because only single date imagery was available, a correlation between ground-measured leaf area index (LAI) and reflectance was established. APAR was then calculated from the spatial distribution of LAI and Beer's Law (Landsberg and Waring 1997), and assumed constant throughout the year. The climate data used was the 1 km grid data previously mentioned (section 3.1), and the soils data was from a 1:50,000 scale coverage. The mean daytime humidity data required to calculate the stomatal closure function were not part of the standard climate dataset, and so were obtained from records at a climate station 10 km away. Extrapolation to the site was by a lapse-rate based method (MTCLIM, Hungerford *et al.* 1989). The forest physiological data used were representative of evergreen broadleaved forest (Landsberg, 2000), and stand age was estimated on the basis of local knowledge. Although it is not possible to properly verify the spatial patterns of NPP predicted, good agreement was obtained between the spatially predicted annual NPP at the locations of the three sites in the area where measurements have been made (Plate 2.b.): agreement within 10% of measured values, or within 15% if stand ages were set simply as either young or medium depending on appearance of the woody vegetation in the SPOT imagery. Differences of this magnitude approach those present in the measured data.

Light use efficiency models of carbon production continue to be improved, with recent advances (Landsberg 2000) that can be expected to overcome some of the limitations seen in early spatial applications where biased, yet well-correlated, predictions resulted (Coops *et al.* 1999, Coops and Waring 2000). At present, it seems that the accuracy of spatial climate and soil layers, especially in topographically complex environments, probably provides a greater limit to the wider use of process-based modelling for estimation of carbon production than does any fundamental problem with the models themselves. Estimating stand age also remains something of a challenge, although for at least plantation forests age is generally available from forest inventory datasets.

3.2.1.   *Predicting Vegetation Species Composition Dynamics*

Vegetation species composition and successional change play a key role in ecosystem function and viability. However, variations in forest composition during growth cycles, or in response to rapid changes in disturbance, competition, or climate, are difficult to represent using statistical models. Instead, a process-based approach is more suited to analysis of the dynamics of vegetation growth and compositional change. Forest gap simulation models have been developed to predict the long-term dynamics of forest response to such things as disease, animal browse, pollution, and large-scale natural or anthropogenic disturbance (see, Shugart and Smith 1996, and references therein). The models are based on the simplifying assumption that vegetation succession is controlled primarily by competition between individual trees for available light and nutrient resources. The LINKAGES model developed by Pastor and Post (1985, 1996) is perhaps the best known of these models, although there are many others.

Forest gap dynamics simulation models are becoming important because for many forests disturbance is becoming more intensive, and the current species compositional mix can not necessarily be assumed to represent steady-state conditions. Given the lifetimes of many forest species, the effect of disturbance may be to induce changes in composition and structure that only become evident over many hundreds of years. The ability to predict the likely long-term outcomes of environmental change, or management intervention, on forest composition is therefore clearly an important part of considering options for environmental policy and planning. In this regard, these comparatively simple growth and competition dynamics models are beginning to produce some remarkable results. For example, parameterising the LINKAGES model for New Zealand conditions results in the prediction of patterns of forest composition and succession that agree very well with those observed at relatively undisturbed sites, for climates that cover the full range from semi-tropical to alpine montane (Hall and Hollinger 2000). Furthermore, for the sites tested, the model does not establish any species where it does not naturally occur, even though canopy gaps are initialised with a complete set of the 75 most common indigenous tree species found throughout the entire country (Hall and Hollinger 2000). The model has also provided insight into the long term changes in vegetation composition that occur in response to major disturbance events many centuries in the past.

To give acceptable predictions, forest gap models do require species-specific information on the effects of light and nutrient availability on growth and mortality, and these data may not always be available. Nonetheless, given the apparent explanatory power of these models, and their ability to represent dynamic systems, it would seem that establishing these parameters should be given priority. At present, application of those models initialised with comprehensive suites of species has been at the plot-scale, based on relatively good climate and soils data. It remains to be established whether spatial climate and soil datasets are sufficiently accurate to allow these models to predict the detailed patterns of vegetation species composition and variation actually observed within complex topography, at landscape scales.

3.3. Knowledge-based Models

Despite the advances outlined above, there remain ecological issues for which we do not have large quantitative datasets with many independent observations to support statistical analysis, nor sufficient knowledge of relevant processes to develop models of species dynamics or ecosystem function. Queries related to the distribution of unusual or rare plant species, for example, provide instances where datasets are sparse, and are probably the result of opportunistic sampling rather than well-designed survey. Often the sample set will be restricted simply to plant presence, rather than the presence/absence or compositional data required for full analyses of potential species distribution. Determining the distribution of animal species poses additional problems, as robust distributional data are generally lacking, and the environmental data most readily available may play only a limited part in determining species distribution. For example, animal behaviours associated with competition and avoidance of disturbance, or complex predator/prey relationships, may be more important predictors than physical habitat characteristics.

For ecosystem monitoring and forecasting associated with these issues, alternative analysis techniques are required: those incorporating or generating knowledge. The range of knowledge-based analysis techniques currently applied to environmental problems are as diverse as the issues being addressed. Rule-based, decision-tree, expert system, machine learning, and neural network techniquesand many variants of thesehave all been applied to the analysis of environmental issues (see, for example, Skidmore *et al.* 1996, and Stockwell and Peters 1999, and references therein). In general, those analysis methods based on decisions or rules are well suited to analysis of sparse or co-dependent data, but explicit knowledge must exist of the relationship between the input data and the set of output solutions/predictions required (Watson 1997). Alternatively, as in the case of expert systems, the necessary knowledge may be elucidated and inferred from a range of trusted opinions, together with whatever other evidence is available. Probabilistic or fuzzy logic methods can be incorporated into these analyses when there is less certainty about any particular rule or decision (e.g. Tucker *et al.* 1997, Bock and Salski 1998, Kampichler *et al.* 2000).

Approaches based on machine learning, primarily genetic algorithms to date, or on neural networks, are also well suited to analysis of many ecological problems. These techniques have the important advantage that the relationship between input data and the output solution/prediction need not be understood. However, both methods require as training data a set of cases that exemplify over the full problem domain the range in output solutions/predictions that occur for various combinations of the input parameters available. The relatively intensive training requirements of these methods also tends to restrict their use to larger datasets, although methods have been proposed recently for application of genetic algorithms to sparse datasets (Stockwell and Peters 1999). Genetic algorithms have the additional useful feature that rules, and thus explanation and justification, are generated and applied during the analysis process. Neural networks, in contrast, function as "black box" analysis systems, providing an answer without justification (Watson 1997).

There is some evidence to suggest that rule-based/decision-tree approaches perform better than the more widely used statistical modelling approaches, even when applied to datasets that statistical analysis has traditionally been considered appropriate for-both in environmental analysis (Skidmore *et al.* 1996), and in other complex fields such as medical and industrial analysis (Watson 1997). This may be due to multi-variate statistical analyses requiring independent variables as input data, whereas in complex systems variables are often co-dependent (Watson 1997, Stockwell and Peters 1999). To date, few studies have compared the results of applying even a limited range of spatial analysis techniques to a common dataset (Stockwell 1992, Odeh *et al.* 1995, Skidmore *et al.* 1996). However, the divergence in the results with analysis method seen in those studies that have been completed suggests that the need is urgent for a more comprehensive comparison of spatial analysis techniques if these techniques are to be used with greater confidence.

3.4. Spatial Information for Ecosystem Analysis and Management: Future Needs

The above examples of ecosystem depiction and modelling raise a number of questions around issues of the quality, availability, and analysis of spatial data that need to be addressed to provide improved information for future environmental management. Beginning with the issue of data quality, it is important to note that as spatial datasets become used more widely for input to complex ecological models, or as part of policy advice, the authors of those datasets exert less and less control over how, and for what purposes, the data are used. Many of those to whom data are delivered will not be sufficiently aware of the limitations imposed by the generalisation processes that inevitably accompany thematic datasets, and especially polygonised datasets-processes used to generate thematic attributes, to place boundaries, to resolve problems associated with overlay operations, and to merge datasets at differing scales. If spatial information is to continue to be a credible part of future ecosystem depiction and modelling, it is critical that more rapid progress be made on including spatial error estimates as an integral part of thematic data supply.

There is no doubt that the past decade has seen a very significant increase in research that provides new methods for investigating error propagation and analysis in particular applications of geographical datasets, and under particular conditions (see, for example, articles in Lowell and Jaton, 1999). However, the inclusion of spatial error estimates as a routine part of data provision remains unusual at present, even though the need for users to be made more aware of the existence and importance of errors in spatial databases was formally raised more than 10 years ago (Openshaw 1989)-and informally over a much longer period. Moreover, Monte-Carlo methods have long been suggested as an operational and pragmatic approach for routinely attaching such error estimates to spatial datasets (Openshaw 1989). Despite this, both the need for increased user awareness, and the requirement for operational use of Monte-Carlo error generation approaches, continue to be recognised as topics requiring urgent international attention (Mowrer 1999). With current computer processing capability now more than sufficient for Monte-Carlo approaches to be implemented as a first approximation error generation method, there seems no excuse for further delay.

The second issue, data availability, relates to almost all ecological analyses at landscape scales being dependent on the existence of accurate climate data. Because ecosystem processes may well depend on climate variables in a non-linear manner, data are ideally required at spatial scales sufficient to represent the detail of ecosystem complexity, that is, essentially at hillslope scales. This is a major challenge. Although relatively reliable methods exist for generating climate layers by surface construction methods (e.g. Busby 1986, Hutchinson and Gessler 1994, Leathwick and Stephens 1998), or by lapse-rate methods based on change in climate with altitude (e.g. Hungerford *et al.* 1989), the spatial accuracy of such layers is presently difficult to establish. In addition, the climate station data on which such layers is based is generally biased towards lower elevation sites, and only rarely samples the montane environments where most land set aside for conservation purposes is located. Consequently, the increased cloudiness typical of many montane environments will not be well represented in many climate station records, adding a particular source of uncertainty to estimation of spatial solar radiation and temperature. Remote sensing has the potential to provide quality assurance data with which to test models of the solar radiation and temperature components of climate surfaces. Measurements of the spatial distribution of rainfall intensity may also become possible in the future (Rosenfeld 1999).

The issue of data availability arises also for those ecological analyses that require accurate information on profile water storage or nutrient status at hillslope scales. This poses an even greater challenge than does obtaining climate data. Internationally, much soils knowledge has been gained for the purposes of intensifying agriculture, and natural areas have received far less attention. In addition, because the cost of field-based soil investigation is high, extrapolation to wider areas is often based on relatively sparse datasets. Advances in geostatistical modelling of soil properties; and quantitative studies linking landscape geomorphology, hydrology, and climate to soil properties; continue to show useful progress (e.g. Odeh et al. 1995, Rogowski 1996, Lagacherie and Holmes 1997, McBratney and Odeh 1997, and references therein). However, it is likely to be some time yet before reasonably detailed surfaces of soil properties exist, with specified accuracy, for areas of complex topography typical of those encountered in many areas of remaining natural vegetation. In the mean time, wider use will probably be made of approaches that segment the landscape into a limited number of hydro-edaphic classes based on landform description and measures of hydraulic flow accumulation (White *et al.* 2000), with measured soil properties assigned to each class. The effort involved in establishing soil properties does, however, limit the application of such approaches. Remote sensing appears to offer limited opportunities to address this information shortfall. At best it is possible to get only the near-surface soil water content from radar data, when there is a significant vegetation cover present. Also, efforts to measure foliar nutrient properties directly using hyperspectral optical data, and so to reduce the need to know soil nutrient conditions, have so far met with limited success (e.g. Johnston and Billow 1996, Fourty and Baret 1998).

The final aspect of data availability relates to the increasing importance being placed on measurement of carbon production in ecosystem studies. As outlined in section 3.2, a common method for doing this requires as a primary data input the amount of incoming photosynthetically active radiation absorbed by the plant canopy (APAR-see equation 1). APAR data for spatial ecological analysis are commonly derived from data obtained from satellites, almost exclusively by correlation with a widely used measure of vegetation green-ness: the normalised difference vegetation index, or NDVI (Tucker 1979). Although it has been demonstrated both theoretically and experimentally that APAR and NDVI are indeed correlated (Sellers *et al.* 1992, Goward *et al.* 1994), the relationship is also a relatively sensitive function of many other parameters that are very seldom accounted for: leaf reflectance and canopy structure, atmospheric conditions, sun and view angles during data acquisition, and the soil background if visible (van Leeuen and Huete 1996, Brown *et al.* 2000, Gobron *et al.* 1999a, Nouvellon *et al.* 2000). Recent work suggests that NDVI-derived estimates of APAR can be in error by up to ±30%, an error that impinges very significantly on estimated carbon accumulation rates (Landsberg and Waring 1997, Gobron *et al.* 1999b). More direct approaches to reliably estimating APAR from remotely sensed data, for a wider range of vegetation types and under a wider range of conditions, have been developed recently and this important development is discussed further in the next section.

In summary, apart from the need for more error analysis and specification in datasets used for ecosystem classification and analysis, there remain significant shortcomings in the availability and reliability of such key spatial datasets as climate, soil, and radiation absorption parameters. Estimation of some of these parameters will benefit from advances in both remote sensing and improved algorithms for processing remotely sensed data. Obtaining finer scale soils data, however, is likely to remain an issue for some time. As well as shortcomings in data, ecosystem analysis would benefit greatly from a well-structured investigation of the merits of alternative analysis techniques, using a common dataset for which sufficient data are available to retain a large independent dataset for verification. Creating and analysing subsets of data that are sparse or poorly structured would further advance our understanding of the usefulness of various analysis and modelling approaches, and particularly help to evaluate the apparent advantages of knowledge-based approaches. These topics should be given more precedence in future research.

4. LANDSCAPE-SCALE INFORMATION FROM REMOTE SENSING: PROGRESS AND CHALLENGES

Improved description of ecosystem function and viability at landscape scales requires both that measures be developed of new and more relevant ecosystem parameters, and that data be available to validate and improve existing spatial databases. We consider here the role remote sensing will play in meeting some of these requirements, through advances in technology and data processing that can be expected to deliver a wider range of ecosystem parameters at enhanced levels of accuracy. It is also important that, as the need for more quantitative environmental data increases, users become better informed of the limits of remote sensing data. These limits become evident particularly when the data are used for monitoring over time, or for extraction of quantitative biophysical information. We briefly review the more important limitations below, together with the good progress now being made to remove these restrictions to wider data use.

4.1. Improved Measures of Vegetation and Biophysical Parameters

In remote sensing, technological development remains fast-paced, with continuing prospects for smaller, lighter, and cheaper satellites and sensors, and consequently for less expensive and more frequent data. Nanosats, satellites weighing in at under 20 kg, are envisaged by the end of the current decade. These are expected to fly in self-organising constellations that autonomously reconfigure to provide differing views and types of images of the earth's surface (NASA 2000a). In terms of information supply for environmental purposes it is useful to group both present and future sensors into three broad classes: coarse to medium spatial resolution optical sensors; fine spatial resolution optical sensors; and radar sensors. These sensor classes represent quite different opportunities for ecological information supply. Table 1 lists some of the more important sensors for the supply of data for terrestrial ecosystem applications that have either been launched recently, or are due for launch over the next few years.

4.1.1 *Coarse and medium spatial resolution optical sensors*

Much effort in space programmes is currently directed towards deploying relatively coarse spatial resolution optical sensors to deliver information every few days at global scales on environmental change on land, in the oceans, and in the atmosphere. The data available from these sensors typically has ground resolutions in the 1 km to 8 km range, with up to 36 spectral bands. (Some sensors also have a limited number of spectral bands with medium spatial resolution, mainly in the 500 m to 250 m range, primarily intended for regional- to local-scale studies.) This new generation of satellites marks a significant advance in the delivery of satellite-based information, in two major ways. First, the sensors have relatively narrow spectral bands carefully placed at wavelengths to enhance sensitivity to particular earth resource phenomena and processes (e.g. Justice *et al.* 1998, Reich *et al.* 1999). This will allow more accurate derivation of biophysical parameters. Second, space agencies have moved to make delivery of derived biophysical parameters an integral part of the suite of standard data products directly available to users.

The MODIS sensor provides a good, though by no means exclusive, example of trends in data quality and supply. For terrestrial targets, the sensor has increased sensitivity to photosynthetic activity, leaf stress and disease, leaf water content, and surface energy exchange (Justice *et al.* 1998, Reich *et al.* 1999, NASA 2000b). Spectral bands have also been placed to maximise sensitivity to aerosols, water vapour, and ozone concentration, all of which are important for terrestrial studies as input parameters to image atmospheric correction and cloud-screening (Cohen and Justice 1999, NASA 2000b). For terrestrial ecosystems, biophysical parameters available as standard global spatial dataset products directly from the satellite operators include: land cover type; vegetation reflectance and vegetation green-ness indices corrected for atmospheric effects, illumination variations, directional reflectance, and the effects of exposed soil; leaf area index; absorbed photosynthetically active radiation (APAR); net primary production; and land surface temperature (Reich *et al.* 1999, NASA 2000c). Up until now, for individual users to derive any one of these products was a significant effort. As experience with these standard products increases, the new generation of coarse spatial resolution sensors should significantly expand the availability of data for ecosystem monitoring and modelling at regional, national, and global scales.

Challenges for information supply, however, remain. The standard data products from these advanced sensors are derived using algorithms whose validity will need to be fully established-although aspects of this are being performed by specialist science teams (Cohen and Justice 1999, Reich *et al.* 1999, Thomlinson 1999, NASA 2000c). In addition, care must be taken in using the algorithms, as they inevitably contain approximations that may not always be met in a given area, or simplifications that result in lower accuracy than expected. Neither is validating the standard biophysical data products at coarser spatial scales of 1 to 8 km a simple undertaking (Cohen and Justice 1999, Reich *et al.* 1999, Thomlinson *et al.* 1999), although a subset of the algorithms for generating standard products will be able to be validated using medium spatial resolution data, at scales of about 250 m (Gobron *et al.* 1999a, NASA 2000c). However, a spatial resolution of even 250 m is still quite large by the standards of normal plot-based ecosystem investigations. There also remain, particularly for pixel sizes of especially 1 km and up but even for 250 m, issues of scaling of non-linear processes that should always be considered when using satellite data to verify output from ecological models, or to provide data for input.

Also remaining in terms of challenges is the ability to discriminate a wider range of vegetation types, as access to even 36 spectral bands of information is likely to afford only a modest increase over that attainable using present sensors that have far fewer bands. This is because many vegetation spectral absorption features are relatively broad-band, and so a relatively small number of spectral bands can capture much of the independent spectral variation related to vegetation type that is present. The use of carefully placed, narrow spectral bands in the newer sensors will help discrimination primarily by being properly centred in areas of higher reflectance dynamic range-such as occurs for variation in chlorophyll absorption around 680 nm. Further discrimination of vegetation type and structure should be possible when multiple-view sensors with a moderate number of spectral bands become available, as both leaf reflectance and vegetation structure have a strong directional reflectance component. The MERIS sensor will be an important sensor in this regard (Martonchik *et al.* 1998, see Table 1). Sensors with a continuous narrow-band spectral coverage from the visible to short-wave infrared region-hyperspectral sensors-should provide further advances in vegetation discrimination, although again not simply in proportion to the number of spectral bands.

Satellite	Sensor	Spectral resolution (number of bands)	Spatial resolution (ground pixel size, m)
Landsat 7	ETM	Vis/nir/tir: 7, pan: 1	vis/nir/tir: 30–60, pan: 15
IKONOS		Vis/nir: 4, pan: 1	vis/nir: 4, pan: 1
EOS TERRA	MODIS		
	ASTER	Vis/nir/tir: 36	250–1000
	MISR	Vis/nir/tir: 14	15–90
		Vis/nir: 4 (9 view directions)	Variable, 240–1000
Quickbird		Vis/nir: 4, pan: 1	vis/nir: 4, pan: 1
ENVISAT	ASAR		
	MERIS	C-band, 2 polarisations.	Variable, 30–100
		Vis/nir: 15	Variable, 250–1000.
ADEOS-II	GLI		
	POLDER	Vis/nir/tir: 36	vis/nir/tir: 250–1000
		Vis/nir: 15 (12 view directions)	vis/nir: 7000
EOS AQUA	MODIS		
		Vis/nir/tir: 36	250–1000
Radarsat 2		C-band, 4 polarisations	Variable: 3–30
SPOT 5	HRV		
		Vis/nir: 4, pan: 1	vis/nir: 10, pan: 2.5
NOAA L, M	AVHRR		
		Vis/nir/tir: 6	vis/nir/tir: 1000–8000

Table 1. Some of the more important satellites/sensors for monitoring terrestrial ecosystems due for launch over the next two to three years. (Many other experimental systems will also be launched, but because of generally restricted access to data are not listed here.) The abbreviations are: vis/nir/tir-visible/near-infrared/thermal infrared, and pan-panchromatic. Near-infrared is considered to include short-wave infrared wavelengths for the purposes of this table. Where "Variable" appears in the spatial resolution column, all bands can be acquired over the range indicated. If a range only is indicated in the spatial resolution column, then the individual bands have resolutions that vary by wavelength (generally larger at thermal infrared wavelengths).

At present, however, it is unclear as to when satellite-based hyperspectral sensors are likely to be routinely providing data, although a number of experimental sensors are planned for launch over the next few years (e.g. the Hyperion sensor, NASA 2000a).

4.1.2   *Fine spatial resolution sensors*

Optical sensors with a high spatial resolution will open up a whole new market in image-based mapping products, that should eventually be considerably less costly than those based on aerial photographs with the same resolution. At present, however, the only commercial civilian satellite with high spatial resolution sensors to reach orbit successfully has been Ikonos-2, carrying a 4 m multispectral sensor, and a 1 m panchromatic sensor. High spatial resolution imagery is expected to find ready use as an orthophoto substitute, and for environmental monitoring where changes in position or area are important. Until image costs reduce, mapping of higher value urban areas is likely to be the major use of high spatial resolution data, with multi-spectral capability providing measures of such variables as urban greenspace and impervious surface area-the latter an important indicator of urban aquatic health (e.g. Schuler 1994, Arnold *et al.* 1996). In the wider environment, such applications as determining the width and location of river channels, the area of sediment deposition or landsliding, or mapping of intricately patterned horticultural lands, are likely to be important uses of such imagery. For ecological studies on forested lands, multi-spectral sensors with 4 m or better spatial resolution will allow the monitoring of incremental changes in position or area, and so provide detection of an important class of ecological alteration: vegetation change that is of low frequency but locally intensive, such as selective defoliation or death of individuals of a tree species within a wider untouched forest cover, or invasion/succession patterns where spectrally distinct individuals gradually encroach on new areas.

The use of high spatial resolution sensors is, however, subject to some significant limitations. Present and near-future satellite-based sensors are somewhat restricted in terms of swath width, although they do have the ability to capture long strips of imagery very rapidly compared with airborne sensors, and the georeferencing of such image strips is considerably simplified compared with equivalent aerial sensor coverage. The revisit time of spaceborne high spatial resolution sensors is also rather long if nadir-viewing imagery is required, so obtaining coverage of a wide area can be a drawn-out process, especially in areas where cloud limits acquisition opportunities. The solution to this problem is to allow off-nadir pointing of the sensor, but this can considerably increase the costs of image georeferencing for other than flat areas. A very high quality digital elevation dataset will be necessary to perform georeferencing of significantly off-nadir viewed imagery if accuracy approaching the image pixel resolution is required. One metre panchromatic imagery represents a significant challenge in this regard, and suitable digital elevation data will rarely exist outside urban areas. Revisit time and coverage problems should diminish somewhat as more sensors achieve orbit, and as new sensors with wider coverage are launched (e.g. on the SPOT-5 satellite). Another area that high spatial resolution imagery can be expected to show limited progress is in general vegetation mapping. This is partially due to the aspects of limited area coverage mentioned above, and partly because present sensors lack the short-wave infrared bands that are so useful for species identification. However, the major limitation is that high spatial resolution satellite images-like much imagery from airborne sensors-resolve many more scene components than previously observed in satellite imagery. Shadows, and differential lighting of the sides and crowns of trees, greatly complicate traditional approaches to automated image analysis based on consistent relationships between image intensity values and target type or biophysical state (e.g. Wilkinson 1996). High spatial resolution imagery should, however, be useful for mapping based on object-based recognition, or on texture analysis, although these techniques remain within the research rather than the operational domain at present.

4.1.3   Radar Sensors

Synthetic aperture radar (SAR) potentially offers some of the more interesting opportunities for the future supply of ecosystem information. Being immune to cloud cover, radar imagery offers increased opportunities for obtaining remotely sensed data in especially tropical areas, where cloud cover often severely limits acquisition opportunities for optical imagery. In terms of measuring vegetation characteristics, the radar signal is sensitive to such properties as physical bulk, stand structure, and dielectric characteristics. These are quite different properties to those contributing to optical imagery, and radar is therefore potentially a useful complementary source of information for many applications. For example, using radar as an added source of information may allow discrimination between plant types that have similar optical spectral properties but differ at the canopy, branch, or stem levels of structural organisation (Freeman and Durden 1998, Pairman *et al.* 1999). Or, by combining both polarimetric and interferometric techniques, retrieval of parameters such as tree height, canopy depth, and understory height can be attempted (Treuhaft *et al.* 1996, Cloude and Papathanassiou, 1998). With carbon production becoming an increasingly important ecosystem parameter, a significant application area for radar is likely to be in the more direct estimation of woody biomass. A number of studies have now demonstrated a relationship exists between above ground biomass and radar backscatter, with the best results obtained when multi-frequency radar imagery is used (e.g. Dobson *et al.* 1995, Harrell *et al.* 1997, Pairman *et al.* 1999).

At present, most of the more advanced applications of radar relate to measurement of vegetation characteristics that rely on data from either airborne or space-shuttle borne sensors, primarily those that are part of the many NASA AIRSAR or Shuttle Imaging Radar missions. In general, only these sensors have the necessary multi-frequency, multi-polarisation capability required to obtain sufficient information to resolve the various features of complex terrestrial targets (Zebker *et al.* 1992, Pairman *et al.* 1999). At present, current satellite-borne sensors are only single frequency, single polarisation instruments, which limits their application for terrestrial studies. Radar data is likely to become much more important when multi-band, multi-polarisation data becomes routinely available-an advance that in terms of information content is similar to that in moving from panchromatic to multi-spectral optical imagery (see Plate 4). Such data will begin to become more widely accessible with the launch of the RADARSAT-2 satellite, scheduled for 2001. Although this satellite will have only a single wavelength, its multi-polarisation capability will greatly enhance the information content of the resultant imagery compared to the existing generation of radar satellites. An example of RADARSAT-2 data synthesised from NASA AIRSAR data is shown in Plate 4.

For SAR to be widely used in terrestrial ecosystem studies, improved models of the interaction between the radar signal and vegetation structural parameters, topography, and surface moisture need to be developed, although steady progress is being made in each of these areas (Oh *et al.*1992, Treuhaft *et al* 1996, Pairman *et al.* 1997, Cloude and Papathanassiou 1998, Schuler 1999). Radar is already contributing important ancillary data for ecological modelling by providing digital elevation data, using both the technique of stereo matching (Leberl, 1998), and the potentially more accurate technique of interferometry (Graham, 1974). With elevation models forming a fundamental spatial data layer for many applications, radar sensors are important for this purpose alone. In the future, radar may also provide information on the spatial intensity and distribution of rainfall. This capability has been recently demonstrated using an experimental active radar sensor that is part of the Tropical Rainfall Measuring Mission (e.g. Rosenfeld 1999). Even sampled measurements of the spatial distribution of rainfall intensity would provide very useful information for evaluating and improving the accuracy of hydrological models, a key component of ecological modelling and forecasting.

4.2 Improved Models for Scene Analysis And Correction

In many instances, application of remote sensing to quantitative monitoring and modelling of terrestrial ecosystems has continued to be hindered by several longstanding problems that introduce variance into the image that is not related to image content: the effects on surface reflectance of variations in atmospheric composition, sun angle, view angle, and topography (e.g. Kriebel 1978, Holben and Justice 1980, Kimes and Sellers 1985, Verstraete *et al.* 1990, Itten and Meyer 1993, Gobron *et al.* 1997, Trotter 1998, Combal *et al.* 2000). Radiometric correction, the process of generating standardised image reflectance values that are independent of these variations, is particularly important when using multi-temporal image sequences to detect ecological change. Without such correction, uncertainty always remains as to whether observed changes are real or simply a function of acquisition conditions and topography. In addition, many advanced processing algorithms for extracting biophysical parameters from imagery require reflectance-standardised imagery as an input dataset. Even in traditional applications like vegetation mapping, that are often performed using single images acquired at near-vertical views, the effect of topography and atmospheric composition becomes important when information is required for areas of steeper hill country and mountain lands (e.g. Itten and Meyer 1993, Trotter 1998). Of the various components of radiometric correction, atmospheric correction is probably the best known and most advanced. As such, we will not discuss it further here, other than to note that the excellent models available for atmospheric correction (Vermote *et al.* 1998) have largely gained operational status now that high quality, narrow-band, global-scale optical spectral data on atmospheric composition is being delivered by the MODIS sensor (Justice *et al.* 1998). Fortunately, as discussed below, advances in processing of satellite imagery over particularly the last ten years mean that the other aspects of radiometric correction can also be performed to a very useful level.

  The topographic effect is most obvious in higher spatial resolution imagery. It appears as an impression of topographic relief, caused by variations in illumination and reflectance with slope and aspect (Plate 3). The effect causes the same target, vegetation say, to appear with quite different brightness values depending on its topographic position. For nadir viewed imagery, a simple empirical correction for the topographic effect is now available (Trotter 1998, Trotter *et al.* 1999), based on the assumption that it arises mainly from variations in illumination, and with only small variations in reflectance due to slope. The correction gives good results for dense vegetation canopies of any type, over a wide range of sun angles, and at both visible and near-infrared wavelengths (Trotter 1998, Trotter *et al.* 1999). Plate 3 shows an example of applying this correction to a Landsat TM image, using terrain data from a digital elevation model derived from 20 m contour data. For off-nadir views, a more sophisticated approach to topographic correction is required, as at some combinations of sun and view angles the surface reflectance can change markedly with slope and aspect, and a simple empirical correction no longer applies (Trotter *et al.* 1999). In these circumstances a correction algorithm must be developed using more complicated reflectance modelling approaches, although quite simple algorithms still result if the effects on scattering of light by different canopy leaf angle orientations are assumed to be negligible (Dymond and Shepherd 1999). Approaches that take account of leaf angle orientation have also been presented recently (Combal *et al.* 2000). Although the latter approach requires information on broad vegetation type, conifers as distinct from broadleaved trees for example, these vegetation types can usually be quite easily delineated in uncorrected imagery. Whether the additional complexity of leaf-angle dependent corrections is warranted, and if so under what image acquisition conditions, is currently being determined.

The variation of surface reflectance with sun and view angle is a problem that becomes very evident when considering time sequential imagery acquired by coarse spatial resolution sensors. The wide swath-width images acquired by these sensors has a reflectance anisotropy caused by both directional scattering and hotspot effects (e.g. Verstraete *et al.* 1990, Gobron *et al.* 1997). The directional scattering effect arises from the tendency of natural surfaces to scatter more radiation back towards the sun than away from it. The hotspot effect is caused by variation in the proportion of shadows seen by an observer, from a minimum when the sun is directly behind an observer and aligned with their line of sight, and increasing as the angle between the sun and the observer diverges. Both effects are functions of canopy three dimensional structure and leaf angle orientation, and combine to give a complex and target dependent pattern of reflectance as a function of view direction in relation to solar geometry. Modelling, and so correcting, for this dependence is not a simple process, ideally requiring significant information on the structural and reflectance characteristics of the particular surface being considered. This degree of information is often not available because identifying the surface type or its biophysical state is the usually the objective of performing the imaging in the first place. Nonetheless, significant improvements in image standardisation can be achieved under a wide range of image acquisition conditions by basing corrections on the surface reflectance properties of generalised land cover classes (Verstraete 1990, Roujean *et al.* 1992, Lewis 1995, Li *et al.* 1995, Gobron *et al.* 1997, Dymond and Qi 1997). These classes can be delineated either by pre-classifying the image before standardisation, or by using available land cover databases for such delineation.

When standardised reflectance datasets are not available, users may choose to implement standardisation approaches themselves, based on published reflectance models. However, not all users will have the necessary experience, nor the necessary computing resources, to undertake such work. An alternative approach has recently become available that, although based on sophisticated reflectance modelling, can be simply implemented to derive more accurate datasets of biophysical parameters useful for ecosystem monitoring and modelling (Govaerts *et al.* 1999). Moreover, it does not require that surface reflectance and observation conditions be specified. The approach uses reflectance modelling to simulate the amount of light (radiance) that would be observed by a sensor in orbit for a given value of the biophysical parameter of interest. The simulation is repeated for the range in both surface type-and so variation in value of the particular biophysical parameter- and observation conditions likely to be encountered. A simple best-fit function, termed an optimised spectral index (Govaerts *et al.* 1999), is then fitted to the resultant data scatter-plot of data describing the relationship between the required biophysical parameter and the radiance observed by the sensor in those spectral bands sensitive to the parameter of interest. So far, this approach has been applied only to derivation of APAR (Gobron *et al.* 1999a), for the MERIS, SeaWIFS, and GLI sensors (Gobron *et al.* 1999a, Verstraete 2000), although there is nothing in principle to prevent its application to deriving a wider range of parameters. It is already apparent that the optimised spectral index approach provides much improved estimates of APAR over the more usual but inherently unsatisfactory methods of estimating APAR by correlation with NDVI (Gobron *et al.* 1999b).

Continued efforts to incorporate standardised reflectance processing into production of image datasets can be expected from satellite operators, if for no other reason than to gain commercial advantage. However, it will be some time before such datasets are routinely produced for all sensors. In the mean time, users should be very aware of the limitations of using imagery with non-standardised reflectance, especially when deriving information from wide swath width or multi-temporal images. When estimates of biophysical parameters must be made from non-standardised imagery, consideration should always be given to adopting optimised spectral indices for parameter estimation (Gobron *et al.* 1999a, Govaerts *et al.* 1999), as these become available for particular parameters of interest.

5. Conclusions

Environmental policy and management are increasingly expected to deliver a careful balance between the use of ecosystems for commodity supply and socio-economic development, and the continuity of these same ecosystems as viable suppliers of life-support services. This can not be attempted with any certainty without an improved ability both to forecast, and to monitor, the impacts of policy and management on ecosystem state and performance at landscape scales. Informed action requires information, and at present much of the spatial information in natural resource databases is not well suited to characterising ecosystems in terms of function and viability, being oriented much more to measures of quantity and location. The challenge for spatial information providers is therefore to devise new approaches to obtaining more relevant information, especially that which can be related to the body of knowledge on processes that determine ecosystem state and performance.

We have outlined here the opportunities for improving the accuracy and relevance of spatial information for environmental analysis and decision making. A key step is considered to be the development of better links between those landscape-level parameters that are observable using remote sensing, and the knowledge and measures of ecosystem state, function, and viability that have been developed largely in plot-based investigations. We have shown how ecosystem models can provide such a link, with input from both remote sensing and spatial databases of climate and soil parameters. Statistical, process-based, and knowledge-based approaches to modelling can all be used, depending particularly on the degree of independence among the input data, the size and representativeness of the input dataset, and the degree to which relationships between input variables and output states are understood. Although it is apparent that much progress has been made in ecosystem classification, modelling, and forecasting, it is also apparent that more attention must be given to removing the shortcomings in the scale and accuracy of spatial data, and also to the specification of spatial error-particularly for the climate and soils spatial datasets that underpin ecological analysis at landscape. Also required is more investigation of the advantages and limitations of the various analysis and modelling approaches applied to ecological datasets, given the apparent divergence in results when differing analysis methods are applied to the same dataset.

We have also examined here the opportunities for meeting at least part of the spatial information shortfall through advances in remote sensing. New coarser spatial resolution sensors such as MODIS will provide not only data that has enhanced sensitivity to biophysical phenomena, but also a suite of standard products that deliver spatial datasets of key biophysical parameters direct to the user. Information to improve estimates of spatial climate parameters will also come from such advanced sensors as MODIS, GLI, ASTER, and MERIS. In addition, advances in the processing of image data have reached the stage where several problems that have significantly hindered the wider application of remote sensing can now be largely removed: directional reflectance and illumination effects, the effect of topography, and correction for atmospheric effects. Turning to fine spatial resolution sensors, we are confronted by a whole new range of opportunities. Image-based maps and aerial orthophoto substitution are likely to be the major applications of these sensors until image costs decline somewhat. General environmental monitoring applications, where small changes in the area or position of spectrally obvious targets must be determined, will then become feasible. Radar data too present a range of new opportunities, although present applications of radar remain limited by the data available from present spaceborne sensors. However, data from airborne sensors indicates that applications related to estimation of forest biomass seem likely, as well as discrimination of plant species that are spectrally similar but structurally distinct. Measurement of spatial rainfall intensity, an important contribution to hydrological modelling, also appears to be a future possibility. In addition, radar already provides information that can be used to derive digital elevation data, an important ancillary dataset for ecological modelling.

Overall, by linking spatial information and ecosystem modelling, much can be achieved towards optimised decision-making in ecosystem conservation, restoration, and utilisation, at landscape scales. Optimised environmental management is becoming a global issue, one whose success will increasingly be monitored through the mechanisms of international agreements, protocols, and reporting. As these mechanisms are ratified and gain legally binding status, we will see further demands, and no doubt changing priorities, for a wider range of spatial information strongly linked to ecological observations and knowledge. For the spatial information community, these are challenging times, but the opportunities for innovation have never been greater.

Acknowledgements

The authors would like to thank Graham Hall and Stephen McNeill for contributing, respectively, information on process-based modelling of plant species distribution and on future satellites and sensors. The preparation of this paper was funded by the New Zealand Foundation for Research, Science, and Technology under contracts C09X006 and C09X0015.

References

Adams, J.M. and Woodward, F.I., 1989, Patterns in tree species richness as a test of the glacial extinction hypothesis, *Nature*, 339, 699–701.

Arnold, A., Arnold, C.L. (Jnr.) and Gibbons, C.J., 1996, Impervious surface coverage: the emergence of a key environmental indicator, *Journal of the American Planning Association*, 62, 243–258.

Aspinall, R. and Veitch, N., 1993, Habitat mapping from satellite imagery and wildlife survey using a Bayesian modelling procedure in a GIS, *Photogrammetric Engineering and Remote Sensing*, 59, 537–543.

Austin, M.P. and Cunningham, R.B., 1981, Observational analysis of environmental gradients, *Proceedings of the Ecological Society of Australia*, 11, 109–119.

Austin, M.P., Nicholls, A.O., and Margules. C.R., 1990, Measurement of the realized qualitative niche: environmental niches of five *Eucalyptus* species, *Ecological Monographs*, 60, 161–177.

Brown, L., Chen, J.M., LeBlanc, S.G. and Cihlar, J., 2000, A shortwave infrared modification to the simple ratio for LAI retrieval in boreal forests: an image and model analysis, *Remote Sensing of Environment*, 71, 16–25.

Bock W. and Salski, A., 1998, A fuzzy knowledge-based model of population dynamics of the yellow-necked mouse (*apodemus flavicollis*) in a beech forest, *Ecological Modelling*, 108, 155–161.

Busby, J.R., 1986, A bioclimatic analysis of *Nothofagus cunninghamia* in south eastern Australia, *Australian Journal of Ecology*, 11, 1–7.

Cloude, S.R. and Papathanassiou, K.P., 1998, Polarimetric SAR interferometry, *IEEE Transactions on Geoscience and Remote Sensing*, 36, 1551–1565.

Cohen, W.B. and Justice, C.O., 1999, Validating MODIS terrestrial ecology products: linking *in situ* and satellite measurements, *Remote Sensing of Environment*, 70, 1–3.

Combal, B. Isaka, H., and Trotter, C.M., 2000, Extending a turbid medium BRDF model to allow sloping terrain with a vertical plant stand, *IEEE Transactions on Geoscience and Remote Sensing*, 38, 798–810.

Coops, N.C. and Waring, R.H., 2000, Estimating maximum potential site productivity and site water stress of the Eastern Siskiyous using 3-PGS, *Canadian Journal of Forest Research (in press)*.

Coops, N.C., Waring, R.H. and Landsberg, J.J., 1998, The development of a physiological model (3-PGS) to predict forest productivity using satellite data, in: Nabuurs, Nuutinen, Bartelink, and Korhonen (Eds), *Forest Scenario Modelling for Ecosystem Management at the Landscape Level*, EFI Proceedings, 19, 173–191.

Coops, N.C., Waring, R.H., and Landsberg, J.J., 1999, Estimation of potential forest productivity across the Oregon transect using satellite data and monthly weather records, *International Journal of Remote Sensing (in press)*.

Daly, G.C., 1997, Introduction: what are ecosystem services?, in Daly, G. (Ed.), *Nature's Services*, Island Press, USA.

Dymond, J.R. and Shepherd, J.D., 1999, Correction of the topographic effect in remote sensing, *IEEE Transactions on Geoscience and Remote Sensing*, 37, 2618–2620.

Dymond, J.R. and Qi, J., 1997, Reflection of visible light from a dense vegetation canopy-a physical model, *Agricultural and Forest Meteorology*, 86, 143–155.

Ehrlich, P. and Ehrlich, A., 1970, *Population, Resources, Environment: Issues In Human Ecology*, W.H Freeman, San Francisco, USA.

Farina, A., 1998, *Principles and Methods in Landscape Ecology*, Chapman and Hall, London.

Freeman, A. and Durden, S.L., 1998, A Three-Component Scattering Model for Polarimetric SAR Data, *IEEE Transactions on Geoscience and Remote Sensing*, 36, 963–973.

Fourty, T.H. and Baret, F., 1998, On spectral estimates of fresh leaf biochemistry, *International Journal of Remote Sensing*, 19, 1283–1297.

Gobron, N., Pinty, B., Verstraete, M.M. and Govaerts, Y., 1999a, The MERIS vegetation Global Vegetation Index (MGVI): description and preliminary application, *International Journal of Remote Sensing*, 20, 1917–1927.

Gobron, N., Pinty, B., Verstraete, M.M. and Mélin, F., 1999b, Development of a vegetation index optimised for the SeaWiFS instrument, *Algorithm Theoretical Basis Document*, JRC Publication EUR 18976 IN, Space Applications Institute, EC JRC, Ispra, Italy.

Gobron, N., Pinty, B., Verstraete, M.M., and Govaerts, Y., 1997, A semidiscrete model for the scattering of light by vegetation, *Journal of Geophysical Research*, 102, 9431–9446.

Govaerts, Y., Verstraete, M.M., Pinty, B. and Gobron, N., 1999, Designing optimal spectral indices: a feasibility and proof of concept study, *International Journal of remote Sensing*, 20, 1853–1873.

Goward, S.N., Waring, R.H., Dye, D.G., and Yang, J., 1994, Ecological remote sensing at OTTER: satellite macroscale observations, *Ecological Applications*, 4, 322–343.

Gower, S.T., Kucharik, C.J. and Norman, J.M., 1999, Direct and indirect estimation of leaf area index, f_{APAR}, and net primary production of terrestrial ecosystems, *Remote Sensing of Environment*, 70, 29–51.

Graham, L.C., 1974, Synthetic interferometer radar for topographic mapping, Proceedings of the IEEE, 62, 763–768.

Hall, G.M.J. and Hollinger, D.Y., 2000, Simulating New Zealand forest dynamics with a generalised temperate forest gap model, *Ecological Applications*, 10, 115–130.

Harrell, P.A., Kasischke, E.S., Bourgeau-Chavez, L.L., Haney, E.M. and Christenson, N.L. Jr., 1997, Evaluation of approaches to estimating above ground biomass in southern pine forests using SIR-C data, *Remote Sensing of the Environment*, 59, 223–233.

Hastie, T. and Tibshirani, R.J., 1990, *Generalised additive models*, Chapman and hall, London.

Hillbert, D.W. and van den Muyzeberg, J., 1999, Using an artificial neural network to characterise the relative suitability of environments for forest types in a complex tropical vegetation mosaic, *Diversity and Distributions*, 5, 263–274.

Holben, B.N. and Justice, C.O., 1980, The topographic effect on the spectral response of nadir pointing sensors, *Photogrammetric engineering and Remote Sensing*, 46, 1191–1200.

Hungerford, R.D., Nemani, R.R., Running, S.W. and Couglan, J.C., 1989, *MTCLIM: a Mountain Microclimate Simulation Model*, USDA Forest Service Research Paper INT-14, Ogden, Utah.

Hutchinson, M.F. and Gessler, P.E., 1994, Splines-more than just a smooth interpolator, *Geoderma*, 62, 45–67.

IGBP, 1998, IGBP Terrestrial Carbon Working Group-The terrestrial carbon cycle: implications for the Kyoto Protocol, *Science*, 280, 1393–1394.

IPCC, 1995, *Intergovernmental Panel on Climate Change: Synthesis Report*, World Meteorological Organisation and Cambridge University Press, Geneva, Switzerland.

Itten, K.I. and Meyer, P., 1993, Geometric and radiometric correction of Landsat TM data of mountainous forested regions, *IEEE Transactions on Geoscience and Remote Sensing*, 31, 764–770.

Johnston, L.F. and Billow, C.R., 1996, Spectroscopic estimation of total nitrogen concentration in Douglas-fir foliage, *International Journal of Remote Sensing*, 17, 489–500.

Justice, C.O., Vermote, E., Townsend, J.R.G., DeFries, R., Roy, D.P., *et al.*, 1998, The Moderate Resolution Imaging Spectrometer (MODIS): land remote sensing for global change research, *IEEE Transactions on Geoscience and remote Sensing*, 36, 1228–1249.

Kampichler C., Barthel J. and Wieland R., 2000, Species density of foliage-dwelling spiders in field margins: a simple, fuzzy rule-based model, *Ecological Modelling*, 129, 87–99.

Karveia, P., 1994, Space: the final frontier for ecological theory, *Ecologyt*, 95, 1–10.

Kimes, D.S. and Sellers, P.J., 1985, Inferring hemispherical reflectance of four vegetated surfaces, *Remote Sensing of Environment*, 18, 205–223.

Knyazikhin, Y., Martonchil, J.V., Mynenei, R.B., Diner, D.J., and Running, S.W., 1998, Synergistic algorithm for estimating vegetation canopy leaf area index and fraction of photosynthetically active radiation from MODIS and MISR data, *Journal of Geophysical Research*, 103, 32257–32275.

Kriebel, K.T., 1978, Measured spectral bidirectional reflection properties of four vegetated surfaces, *Applied Optics*, 17, 253–259.

Lagacherie, P. and Holmes, S., 1997, Addressing geographical errors in a classification tree for soil unit prediction, *International Journal of Geographic Information Systems*, 11, 183–198.

Landsberg, J.J., 2000, Personal communication.

Landsberg, J.J. and Gower, S.T., 1997, *Applications of Physiological Ecology to Forest Production*, Academic Press, SanDiego.

Landsberg, J.J. and Waring, R.H., 1997, A generalised model of forest productivity using simplified concepts of radiation-use efficiency, carbon balance, and partitioning, *Forest Ecology and Management*, 95, 209–228.

Landsberg, J.J., Johnsen, K.H., Albaugh, T.J., Allen, H.L., and McKeand, S.E., 2000, Applying 3-PG, a simple process-based model designed to produce practical results, to data from loblolly pine experiments, *Forest Science (in press)*.

Law, B.E., Baldocchi, D.D., and Anthoni, P.M., 1999, Below-canopy and soil CO_2 fluxes in Ponderosa pine forest, *Agricultural Forest Meteorology*, 94, 171–188.

Leathwick, J.R., 1995, Climatic relationships of some New Zealand forest tree species, *Journal of Vegetation Science*, 6, 237–248.

Leathwick, J.R., 1998, Are New Zealand's Nothofagus species in equilibrium with their environment?, *Journal of Vegetation Science*, 9, 719–732.

Leathwick, J.R., 2000, New Zealand potential forest pattern predicted from current species-environmental relationships, *New Zealand Journal of Botany* (in press).

Leathwick, J.R. and Stephens, R.T.T., 1998, *Climate surfaces for New Zealand*, Landcare Research Contract Report LC9798/126, 19 pp, Landcare Research, Lincoln, New Zealand.

Leathwick, J.R. and Austin, M.P., 2000, Competitive interactions between forest tree species in New Zealand's old-growth forests, *Ecology (in press)*.

Leathwick, J.R., Burns, B.R. and Clarkson, B.D., 1998, Environmental correlates of tree alpha-diversity in New Zealand primary forests, *Ecography*, 21, 235–246.

Leathwick, J.R., Whitehead, D. and McLeod, M., 1996, Predicting changes in the composition of New Zealand's indigenous forests in response to global warming: a modelling approach, *Environmental Software*, 11, 81–90.

Leberl, F.W., 1998, Radargrammetry, in *Principles And Applications of Imaging Radar*, Henderson, F.M. and Lewis, A.J. (Eds), John Wiley & Sons, New York, 183–251.

Levin, S.A., 1992, The problem of pattern and scale in ecology, *Ecology*, 73, 1943–1967.

Lewis, P., 1995, The utility of kernel-driven BRDF models in global BRDF and albedo studies, *Proceedings of the International Geoscience and Remote Sensing Symposium*, 1186–1187.

Li, X., Strahler, A., and Woodcock, C.E., 1995, A hybrid geometric optical-radiative transfer approach for modelling albedo and directional reflectance of discontinuous canopies, *IEEE Transactions on Geoscience and Remote Sensing*, 2, 466–480.

Lowell, K. and Jaton, A., 1999, *Spatial Accuracy Assessment: Land Information Uncertainty in Natural Resources*, Lowell, K. and Jaton, A (Eds), Ann Arbor Press, Michigan.

Martonchik, J.V., Diner, D.J., Pinty, B., Verstraete, M.M., Myneni, R.B., Knyazikhin, Y. and Gordon, H.R., 1998, Determination of land and ocean reflective, radiative, and biophysical properties using multi-angle imaging, *IEEE Transactions on Geoscience and Remote Sensing*, 36, 1266–1281.

McBratney, A.B. and Odeh, I.O.A., 1997, Application of fuzzy sets in soil science: fuzzy logic, fuzzy measurements, and fuzzy decisions, *Geoderma*, 77, 85–113.

Monteith, J.L., 1977, Climate and the efficiency of crop production in Britain, *Philosophical Transactions of the Royal Society of London, Series B*, 281, 277–294.

Mowrer, H.T., 1999, Accuracy (re)assurance: selling uncertainty assessment to the uncertain, in *Spatial Accuracy Assessment: Land Information Uncertainty in Natural Resources*, Lowell, K. and Jaton, A. (Eds), Ann Arbor Press, Michigan, 3–10.

Margules, C.R. and Pressey, R.L., 2000, Systematic conservation planning. *Nature*, 405, 243–253.

NASA, 2000a, *The New Millennium Project*-http://nmp.jpl.nasa.gov

NASA, 2000b, *The MODIS Sensor*-http://ltpwww.gsfc.nasa.gov/MODIS/ MODIS.html

NASA, 2000c, *MODIS Algorithm Theoretical Basis Documents*-http://modarch.gsfc. nasa.gov/MODIS/ATBD/atbd.html

Nouvellon, Y., Seen, D.L., Rambal, S., Begue, A., Moran, M.S., Kerr, Y. and Qi, J., 2000, Time course of radiation use efficiency in a shortgrass ecosystem: consequences for remotely sensed estimation of primary production, *Remote Sensing of Environment*, 71, 43–55.

Odeh, I.O.A., McBratney, A.B. and Chittleborough., 1995, Further results on prediction of soil properties from terrain attributes: heterotopic cokridging and regression kridging, *Geoderma*, 67, 215–226.

Oh, Y., Sarabandi, K. and Ulaby, F.T., 1992, An empirical model and an inversion technique for radar scattering from bare soil surfaces, *IEEE Transactions on Geoscience and Remote Sensing*, 30, 370–382.

Openshaw, 1989, Learning to live with errors in spatial databases, *in Accuracy of Spatial Databases*, Goodchild, M. and Gopal, S. (Eds), Taylor and Francis, London, 263–276.

Pairman, D. Belliss, S.E. and McNeill, S.J., 1997, Terrain influences on SAR backscatter around Mt Taranaki, New Zealand, *IEEE Transactions on Geoscience and Remote Sensing*, 35, 924–932.

Pairman, D., McNeill, S., Scott, N., Belliss, S. 1999, Vegetation identification and biomass estimation using AIRSAR data, *Geocarto International*, 14, 67–75.

Pastor, J. and Post, W.M., 1985, *Development of a Linked Forest Productivity-soil Process Model*, Oak Ridge National Laboratory, ORNL/TM-9519, Oak Ridge, Tennessee, USA.

Pastor, J. and Post, W.M., 1996, LINKAGES: an individual-based forest ecosystem model, *Climatic Change*, 34, 253–261.

Pereira, J.M.C. and Itami, R.M., 1991, GIS-based habitat modelling using logistic multiple regression: a study of the Mt Graham red squirrel, *Photogrammetric Engineering and Remote Sensing*, 57, 1475–1486.

Prince, S.D. and Goward, S.N., 1995, Global primary production: a remote sensing approach, *Journal of Biogeography*, 22, 815–835.

Reich, P.B., Turner, D.P. and Bolstad, P., 1999, An approach to spatially distributed modelling of net primary production (NPP) at the landscape scale and its application in validation of EOS NPP products, *Remote Sensing of Environment*, 70, 69–81.

Rogowski. A.S., 1996, Quantifying soil variability in GIS applications: II Spatial distribution of soil properties, *International Journal of Geographic Information Systems*, 10, 455–475.

Rosenfeld, D., 1999, TRMM observed first direct evidence of smoke from forest fires inhibiting rainfall, *Geophysical Research Letters*, 26, 3105–3108.

Roujean, J-L., LeRoy, M. and Deschamps, P.Y., 1992, A bidirectional reflectance model of the Earth's surface for the correction of remote sensing data, *Journal of Geophysical Research*, 97, 20455–20468.

Running, S.W. and Gower, S.T., 1991, FOREST-BGC, a general model of forest ecosystem processes for regional applications. II. Dynamic carbon allocation and nitrogen budgets, *Tree Physiology*, 9, 147–160.

Running, S.W. and Hunt, E.R. Jr., 1993, Generalisation of a forest ecosystems process model for other biomes, BIOME-BGC, and an application for global scale models, in *Scaling Processes Between Leaf and Landscape Levels*, Ehleringer and Field (Eds), Academic press, San Diego, 141–158.

Rutherford, M.C., Powrie, L.W. and Schulze, R.E., 1999, Climate change in conservation areas of South Africa and its potential impact on floristic composition: a first assessment, *Diversity and Distributions*, 5, 253–262.

SCEP, 1970, *Study of Critical Environmental Problems (SCEP), Mans Impact on the Global Environment*, MIT Press, Cambridge, Massachusetts, USA.

Scepan, J. Davis, F. and Blum, L.L., 1987, A geographic information system for managing California Condor habitat, *Proceedings of GIS/LIS '87* (Bethesda, Maryland; American Society for Photogrammetry and Remote Sensing), 476–486.

Schuler, T.R., 1994, The importance of imperviousness, *Watershed Protection Techniques*, 1, 100–103.

Schuler, D.L., Lee, J-S. and Ainsworth, T.L., 1999, Compensation of terrain azimuthal slope effects in geophysical parameter studies using polarimetric SAR data, *Remote Sensing of Environment*, 69,139–155.

Sellers, P.J., Berry, J.A., Collatz, G.J. Field, C.B., and Hall, F.G., 1992, Canopy reflectance, photosynthesis and transpiration. III. A reanalysis using improved leaf models and a new canopy integration scheme, *Remote Sensing of Environment*, 42, 187–216.

Shugart, H.H. and Smith, T.M., 1996, A review of forest patch models and their application to global change research, *Climatic Change*, 31, 131–153.

Skidmore, A.K., Gauld, A. and Walker, P., 1996, Classification of kangaroo habitat distribution using three GIS models, *International Journal of Geographical Information Systems*, 10, 441–454.

Sperry, R.O., Katul, G.G., Pataki, D.E., Ewers, B.E., Philips, N. and Schäfer, K.V.R., 1999, Survey and synthesis of intra- and interspecific variation in stomatal sensitivity to vapour pressure deficit, *Plant, Cell and Environment*, 22, 1515–1526.

Stockwell, D.R.B., 1992, *Machine Learning and the Problem of Prediction and Explanation in Ecological Modelling*, Doctoral thesis, Australia National University, Canberra, Australia.

Stockwell, D. and Peters, D., 1999, The GARP modelling system: problems and solutions to spatial prediction, *International Journal of Geographical Information Systems*, 13, 143–158.

Stohlgren, T.J., Binkley, D., Chong, G.W., Kalkhan, M.A., Schell, L.D., Bull, K.A., Otusuki, Y., Newnman, G., Bashkin, M. and Son, Y. (1999, Exotic plant species invade hot spots of native plant diversity, *Ecological Monographs*, 69, 25–46.

Thomlinson, J.R., Bolstad, P.V. and Cohen, W.B., 1999, Coordinating methodologies for scaling landcover classifications from site-specific to global: steps towards validating global map products, *Remote Sensing of Environment*, 70, 16–28.

Trotter, C.M., 1998, Characterising the topographic effect at red wavelengths using a juvenile conifer canopy, *International Journal of Remote Sensing*, 19, 2215–2221.

Trotter, C.M., Combal, B. and Pinkney, E.J., 1999, Measuring and modelling the topographic effect at red and near-infrared wavelengths for juvenile trees on an orientable platform, *Proceedings of the Remote Sensing Society Conference*, pp. 349–356, Cardiff, UK.

Treuhaft, R.N., Madsen, S.N., Moghaddam, M. and van Zyl, J.J., 1996, Vegetation characteristics and underlying topography from interferometric radar, *Radio Science*, 31, 1449–1485.

Tucker, C.J. 1979: Red and photographic infrared linear combinations for monitoring vegetation, *Remote Sensing of Environment,* 8, 127-150..

Tucker K., Rushton, S.P., Sanderson, R.A., Martin E.B. and Blaiklock J., 1997, Modelling bird distributions – a combined GIS and Bayesian rule-based approach, *Landscape Ecology*, 12, 77–93.

van Leeuwen, W.J.D. and Huete, A.R., 1996, Effects of standing litter on the biophysical interpretation of plant canopies with spectral indices, *Remote Sensing of Environment*, 55, 123–138.

Vermote, E., Tanré, D., Dueze, J.L., Herman, M. and Mocrette, J.J., 1998, Second simulation of the satellite signal in the solar spectrum: an overview, *IEEE Transactions on Geoscience and Remote Sensing*, 35, 675–686.

Verstraete, M.M., 2000, *Personal Communication*, Space Applications Institute, EC JRC, Ispra, Italy.

Verstraete, M.M., Pinty, B. and Dickenson, 1990, A physical model of the bidirectional reflectance of vegetation canopies, *Journal of Geophysical Research*, 95, 111765–11775.

Watson, I.D., 1997, *Applying Case Based Reasoning: Techniques for Enterprise Systems*, Morgan Kaufmann, San Francisco.

White, J.D., Running, S.W., Thornton, P.E., Keane, R.E., Ryan, K.C., Fagre, D.B. and Key, C.H., 2000, Assessing simulated ecosystem processes for climate variability research at Glacier National park, USA, *Ecological Applications (submitted)*.

Waring, R.H. and Pitman, G.B., 1985, Modifying lodgepole pine stands to change susceptibility to mountain pine beetle attack, Ecology, 66, 889–897.

Waring, R.H., and Running, S.W., 1998, *Forest Ecosystems (2nd Edition)*, Academic Press, San Diego.

Walker, P.A. and Young, M.D. (1997) Using integrated economic and ecological information to improve government policy, *International Journal of Geographical Information Systems*, 11, 619–632.

Wilkinson, G.G., 1996, A review of current issues in the integration of GIS and remote sensing, *International Journal of Geographical Information*

Yee, T.W. and Mitchell, N.D., 1991, Generalised additive models in plant ecology, *Journal of Vegetation Science*, 2, 587–602.

Zebker, H.A., Madsen, S.N., Martin, J., Wheeler, K.B., Miller, T., Lou, Y., Alberti, G., Vetrella, S. and Cucci, A., 1992, The TOPSAR interferometric radar topographic instrument, *IEEE Transactions on Geoscience and Remote Sensing*, 30, 933–940.

2

Methodologies And Approaches For Automated Land Cover Change Detection

A.J.Comber, A. N. R. Law, J. R. Lishman

1. Introduction

The thesis proposed here is that by integrating a wide range of data and knowledge about the drivers of land cover and land cover change, an automated approach to land cover mapping can be developed that is neither region nor application specific.

2. Global Land Cover Mapping Requirement

There is an international requirement to monitor land cover and land cover change. This is reflected by a variety of projects, which have been initiated to collect and analyse land cover data such as the UK Environmental Change Network (ECN, 2000), the European Environment Agency (EEA, 2000) and the International Geosphere-Biosphere Program (IGBP, 2000). Common to all these programmes is the wider definition of "global change" that includes the monitoring of vegetation and changes in land use. The issue being addressed in this paper is how to locate areas of change through the development of computer experts to represent the factors in specific land cover change events and their interaction within an Artificial Intelligence (AI) setting. The application is being developed to monitor land cover change in Scotland, but the conceptual framework within which the task is being approached has implications for monitoring the effects of global environmental change.

2.1. The Land Cover of Scotland survey

The Land Cover of Scotland survey (LCS88) was the first detailed census of the land cover of Scotland and provided baseline information at a scale of 1:25,000. It was instigated in 1987 as a result of increasing concern about the nature and rate of change of land cover in rural Scotland (specifically semi-natural vegetation) and the need for baseline information on which to base future countryside policy and research. LCS88 is a digital dataset derived from the manual interpretation of air photography and the conversion of this land cover information to a GIS (MLURI, 1993). The original choice of aerial photograph interpretation (API) was to enable the mapping of detailed and localised land cover features and has subsequently been recommended for any repeat Land Cover of Scotland (Dunn *et al*, 1995). However LCS88 was a costly and labour intensive exercise that took some 20 people years to complete and consequently there has been considerable interest in automated techniques for updating the land cover information from this baseline (Birnie *et al*, 1996). In a methodical review of the possible approaches for automating air-photo interpretation (API) for any repeat Land Cover of Scotland survey, Horgan *et al* (1997) recommended that a semi-automatic approach be pursued through the use of auxiliary information such as topography, soil maps, and previous classifications.

2.2.   SYMOLAC – a system for monitoring land cover

Any automated approach to land cover monitoring has to be able accommodate auxiliary information such as topography, soil maps, previous classifications, and knowledge of how to apply them. SYMOLAC is a G2-based (Gensym, 2000) prototype artificial intelligence system for detecting land cover change in the interval 1988 to 1995 by linking the spectral analysis of high resolution satellite imagery to spatial interrogation of Geographic Information System via bridges to Visual Numeric's PVWAVE and ESRI's ARC/INFO respectively (Skelsey, 1997). The initial land cover change event that was monitored was the felling of coniferous plantation forestry and this has a number of advantages for prototype development:
-    felled forestry is spectrally distinct event;
-    the number of change events are limited – the class of "Coniferous woodland: plantation" will realistically only change to "Recently felled forestry" in the time period;
-    any changes will occur within the original polygon, so no neighbouring polygons have been considered in the analysis.

In artificial intelligence terms SYMOLAC's novelty lies in its overall approach, rather than in the design of its individual components. SYMOLAC has at its core the notion that the reasoning strategy should be able to adapt to the nature of the problem, a concept named "task orientation" by Skelsey (1997). It is a technique that focuses on the original aims of the study and the most effective way of attaining them using the available resources (rather than demonstrating what can be achieved using a specific technique). A solution strategy emerges dynamically rather than being tailored to suit each task, the data and knowledge available.

3.   The broad aim

SYMOLAC ver1.0 detects one specific land cover change event to prove the concept of design and the task-oriented reasoning strategy. The broad aim of this work is to expand SYMOLAC's knowledge base, through the development of computer experts to represent other land cover transition events. Some of the issues surround the mapping of land cover from remotely sensed imager is introduced in Section 3.2. The effects of biophysical and managerial influences on land cover, and that of their interaction are described in Section

3.3 and an outline description of the computing concepts is given in Section 3.4. Section 3.5 is intended as a background to the ecological knowledge modelling concepts within the paper and the reader is directed to Skelsey (1997) for a complete description of SYMOLAC. Section 3.6 details the work that has been done to expand the monitoring capacity of SYMOLAC and illustrates this with a worked example. Section 3.7 outlines some possible areas for further work.

3.1.1.   Land cover mapping from remote sensed imagery

Land cover is often mapped using remotely sensed data by one of two methods:

1.   Statistical clustering of digital imagery
2.   Manual interpretation

Statistical clustering of high-resolution satellite imagery (supervised or unsupervised classification) is relatively quick, produces moderately accurate results (typically 70-85%) and, due to its digital nature, has resulted in some semi-automated approaches (for example, Salvador *et al*, 2000; Waldemark, 1997; Fuller *et al*, 1994). The limitations of this technique are that it relies on the land cover being mapped to be spectrally distinct in terms of pixel values. This produces good results when broad scale classifications are used (i.e. a limited number of clusters), but fails to delineate between all of the land cover classes with more detailed classification schemes (such as LCS88). These limitations are compounded in a heterogeneous semi-natural environment, where different vegetation communities can intergrade and may not be separated by clearly defined boundaries. The information content of the spectral data may be limited in trying to derive a more detailed classification and can often produce results that are application or region specific (for example, Belward *et al* 1990; Green *et al* 1994; Holmgren and Thuresson, 1998; Macleod and Congalton 1998; Lyon *et al*, 1998). This has lead many workers to identify the need for ancillary information to assist in the classification of remotely sensed imagery (for example, Foody and Hill, 1996; Mattikalli *et al* 1995; Stuckens *et al*, 2000; Wright and Morrice, 1997, Green *et al* 1994; Holmgren and Thuresson, 1998).

Visual interpretation of imagery, typically aerial photographs (AP), relies on the human interpreter being able to delineate between the different land cover classes. The robustness of this method is due to the ability of the human interpreter to derive more detailed contextual information from the AP (topography, tonal and textural differences, soil type, etc.) and the application of local first hand knowledge to the interpretation process. But this technique is inherently subjective and there may be strong differences between interpreters over boundary location, ascribing a parcel of land into a particular land cover class and the way that the minimum mapping unit is treated (Cherrill and McClean, 1995; Burrough, 1986). These differences can be compounded in semi-natural environments and with more detailed classification schemes such as LCS88 (Comber *et al*, 1999). There are other ways that vegetation communities can be delineated and represented such as through the use of fuzzy boundaries, indicating the transition between adjacent land cover classes (Goodchild, 1994). However in the case of the LCS88 survey, conventional cartographic representations were used because of the time implications and the origins of the project. If each polygon boundary was to have its "crispness" evaluated as part of the API process then this would have increased the time to complete the survey. There would also be implications for the validation exercise, which would have to have accounted for variation in boundaries. The origin of the project was to quantify the extent of heather moorland in Scotland, and the use of a conventional cartographic representation, despite all its failings, made this a more transparent exercise for those who had commissioned it. The gross land cover figures produced by LCS88 were similar to those from the Land Cover Map of Great Britain (Fuller *et al*, 1994) and the National Countryside Monitoring Survey (Tudor *et al*, 1994), despite all three surveys having very different methodologies.

In summary, the modelling of land cover mapping from remotely sensed imagery, at the level of detail of LCS88, requires additional information to that which can be derived from the spectral data alone.

3.1.2.   Land Cover Determinants

The vegetation at any point is determined by many factors. Broadly these can be separated into two categories:

- Environmental factors – any particular bio-geographical niche will *tend* to have a particular plant community due to the interplay of different environmental gradients.
   Management practices – the tendency towards the natural land cover is mitigated by decisions on how to manage the land.

The presence of any particular vegetation community can be seen as the interaction between the environmental gradients and the management practice at that point. In a detailed taxonomy (for instance plant - mychorhizal associations), very specific plant and associated root-zone bacteria communities may be found only where there is a specific interaction between the environmental and managerial factors. As the vegetation community definition becomes broader then they may not occupy such a specific position in this multi-dimensional environmental and managerial feature space.

Determining "land cover" is the categorisation of the vegetation into discrete land cover classes. To model the classification process in terms of the different environmental and managerial factors, all the possible positions of a land cover class in this multi-dimensional environmental and managerial feature space must be considered. This can be illustrated by one of the LCS88 single feature classes. Undifferentiated Coarse Grassland was described as being composed of communities that were dominated by either *Nardus stricta* or *Molinia caerulea*. Although they occupy very different positions in environmental feature space (the *Molinia caerulea* prefers very wet conditions and is not in abundance East of the "Molinia Line"), they are both contained within the class of Undifferentiated Coarse Grassland.

This class can also be approached by modifying the management of a particular area. Most of the moors in Scotland that fall below the treeline, would naturally be covered in a variety of woodland types, were there no management of that land. For hundreds of years the moors have been managed either for sport or sheep. Management practice has been to periodically burn parts of the moorland and to adjust stocking densities. Burning removes the heather and creates a niche that the grass species are able to fill before the heather recolonises it after two or three years. If the moor is overburned or overstocked the heather can disappear to be replaced by the grasses. So by increasing the stocking rates of an area the land cover can be modified to a grass community of either *Nardus stricta* or *Molinia caerulea* (LCS88 class Undifferentiated Coarse Grassland) depending on the underlying environmental gradients. That is, there are more than one combination of environmental and managerial factors that will result in the same land cover class.

3.1.3. Artificial Intelligence Techniques

There have been consistent indications of a need to incorporate ancillary data or domain knowledge into the classification process and various researchers have integrated a range of data to improve the accuracy of land cover classification. In turn this has resulted in some semi-automated techniques to improve the classification accuracy by incorporating external data (for example, Smith and Fuller, in press; Huang and Jensen, 1997; Myeni, *et al* 1995). Whilst some of the automated approaches have successfully managed to incorporate a body of knowledge and rules into an automated approach to land cover mapping, these have been for particular sets of mapping constraints and cannot be applied to all land cover classifications. For instance parcel-based approach developed by Smith and Fuller (in press) uses OS LandLine data to improve the per polygon classification.

Where the land cover is discretised by some management boundary (such as a fence) that is represented on the LandLine dataset, this is a very attractive solution. Where the vegetation communities are semi-natural these management divisions do not exist, and the incorporation of environmental knowledge and datasets assists the analysis. Knowledge of the spatial characteristics of land cover features derived from remotely sensed imagery has been used to help automated land cover classifications for specific cover types such as building detection (Zhang, 1999) or woodlands (Blackburn and Milton, 1997). Other automated techniques have been applied to determine broad-scale land cover classes and in a particular area (e.g. Huang and Jensen, 1997; Chavez and MacKinnon, 1994; Kartekeyen *et al* 1995).

The general tendency in automated techniques to land cover has been to apply a body of knowledge to the classification problem in a consistent and pre-determined manner. These approaches to modelling within an artificial intelligence system are through a fixed rule base or knowledge tree. The reasoning process moves through the knowledge base via a forward- or backward-chaining mechanism, and each of the branches in the knowledge base represents the outcome of a decision. This results in automated solutions that are very domain-, process- or region-specific. This is in part due to the dimensionality of the knowledge-base which often apply a set of rules that relate to one aspect of the classification process (e.g. Smits and Annoni, (1999) using texture information; Bruzzone and Prieto (2000) using assumptions about pixel independence), sometimes two (e.g. Gumbricht *et al*, 1996 using geological and elevation data) and always in a pre-determined manner. That is, the order of knowledge application depends on the structure of the knowledge base and the order of rule firing is predetermined by the knowledge tree structure. This results in a degree of cohesion between reasoning, control and representation within the expert system, and this in turn produces fixed, rigid approaches to problem solving.

These sorts of rule-based expert system are suited to applications that have a limited number of factors, evaluated in pre-determined way. They tend to be *method-based*, either ignoring or engineering out real world complexities. However, the determination of land cover change depends upon many factors and the particular ordering of the knowledge, weighting of evidence, and so on will change, depending on the land cover transition being examined, the bio-geographic location and management. The knowledge base would have to represent every possible combination of the different factors involved in determining change for all of the land cover classes. At a local or regional level, with a simple land cover classification, it may be possible to represent and model the land cover transitions in a very complicated decision tree. But the size of such a tree becomes large at national (or international) scales with ecologically complex classification schemes. The knowledge-base would have to represent every possible combination of all the different factors involved in determining change for all land cover classes and it would be too complex and computationally inefficient. This complexity and inefficiency would be compounded in semi-natural environments like Scotland and for detailed classification schemes like LCS88.

3.1.4. SYMOLAC: The Prototype

Symolac ver1.0 (Skelsey, 1997) is a G2 based artificial intelligence system for detecting land cover change. Its principal features are a task oriented approach, a blackboard based model, integration of GIS and remote sensing capability and a prototype application. Below is a description of the components of SYMOLAC, but the reader should be aware that this is limited description of a complex area and that a complete description of the concepts and components of SYMOLAC can be found in Skelsey (1997).

3.5.   Task Orientation

Given that the actual land cover at any point is the result of any one of a number of potential interacting factors, and that the knowledge and reasoning needed to arrive at the solution will also be different, a "task oriented" approach (Skelsey, 1997) to the design was taken. Different land cover transitions require different solutions, and the ethos behind the approach can be illustrated by the way that remotely sensed data is treated. The typical remote sensing approach to land cover mapping takes the position of "how can x be monitored using spectral data?" Information retrieval is modelled to the problem area and to the manipulation of domain knowledge in a specific way. Typically the spectrally distinct cover types are classified, changes identified, and biomasses correlated to vegetation indices. Other more spectrally heterogeneous land covers are less accurately identified. Improvements can be made by fine tuning the analysis, but the results become very instance specific and subjective. The usual conclusions, demonstrated by many examples of investigations using these techniques, however is that they work for some coverages or classes, or areas and not for others (for example, Macleod and Congalton, 1998; Lyon *et al*, 1998). This is supported by the work of Singh (1989) who compared change detection techniques for forestry and memorably stated that "the conclusion is that even in the same environment various techniques may yield different results", the corollary being that the same technique will produce different results in different environments.

The task-oriented approach, in this remote sensing example, would be to ask, "How may spectral data assist in the monitoring of x?". In this sense any remotely sensed imagery would be seen as one of a number of potentially useful datasets that would help to derive a solution to the problem. By fully developing this line of reasoning the task oriented approach sought to:
-   incorporate multi-source data;
-   represent and reason with disparate knowledge;
-   to propagate uncertainty through any analysis;
-   adapt analyses to suit the nature of each task;
-   interface with task suitable software packages;
-   generate detailed explanations of processing and reasoning.

3.6.   Blackboard Based Model

The failings of rule based expert systems are overcome in Symolac by a blackboard based model (see Englemore and Morgan (1988) for a full description of blackboard model concepts, or Skelsey (1997) for a full explanation of its use in SYMOLAC). The aim was to reduce the cohesion between the domain concepts and the reasoning process. The blackboard model inside SYMOLAC has three components (Skelsey, 1997):
-    collection of individual computer "expert";
-    an area of shared memory, the "Blackboard"";
-    a mechanism for monitoring or controlling the application of each expert.
   Objects on the blackboard represent the current state of the solution of the problem and are observed at all times by the experts. Experts check to see if events are "of interest" to them: if so they are invoked. The control component of the blackboard-based model chooses experts to activate at each stage of the reasoning and selects from those that have a state of "interested in". They are chosen dynamically by the controller, which examines the current state of the solution and employs knowledge of the problem domain to order the competing experts' responses. The ordering is not defined by the application, rather it is a process independent of the knowledge base. The blackboard controller changes the focus of attention from one part of the solution to another – a shift from linear "search for solutions" of traditional knowledge based systems to a more co-operative, flexible knowledge assembly (Skelsey, 1997). Application areas suitable for blackboard solutions are where there are different types of knowledge, a variety of datasets that describe aspects of a problem and where data is sparse and of varying quality (Englemore and Morgan, 1988).
   In Symolac this co-operation between experts is done through the use of "goals". Goals are posted on the blackboard by experts, they are requests for knowledge, data, GIS or image processing, etc., and are responded to by other experts. Experts perform many roles: they can represent specific land cover change events (hypothesise changes to a particular LCS88 land cover class), they can be data management experts ("get all the geospatial data for this area") they can schedule other experts (order the response to particular goals) and they can evaluate the evidence in support of any change or null hypothesis. In this way reasoning and knowledge can be represented in different ways by the different experts, even though they do not communicate with each other directly (they use the blackboard).

3.6.1   *The integration of GIS and remote sensing capability via bridges to ESRI's Arc/Info and Visual Numeric's PVWAVE*

Historically the interface between the GIS and remote sensing is functional but weak (Jensen, 1996). Each side suffers formal lack of critical support that could be provided by the other: GIS has a need for timely accurate updates, and remote sensing systems could benefit from highly accurate ancillary information to more usefully extract information from the data (Ehlers *et al*, 1991; Lunetta *et al*, 1991). In Symolac the Arc/Info GIS manages all the spatial data and provides spatial analysis. The image analysis package PV-WAVE provides statistical and analysis functions.

3.6.2 *Prototype application*

These features have been implemented to look for a singular land cover change event – plantation forestry felling – in a test area West of Aberdeen using Landsat TM images from 1988 and 1995. This land cover transition was suitable for the initial development of the prototype for a number of reasons:

- both plantation forestry and felled plantation forestry are spectrally distinct;
- the number of change events are limited – commercial forestry can only change to felling;
- the hypotheses to be evaluated are limited: a full felling event (the whole polygon), a partial felling event (part of the polygon), or the null hypothesis (no change);
- the change will occur within the original plantation forestry polygon (neighbouring polygons do not have be considered in the analysis);
- the expert knowledge was confined to whether the suspected partial felling was large enough, was too close to a track (and therefore *was* a track), and the presence of adjacent felled areas;
- the evidential reasoning was done with the support of the various pieces of evidence for and against the change hypotheses to determine whether land cover change had occurred in this example.

If at any point in the reasoning process a "good enough" solution can be concluded – that is, there is conclusive endorsement for a piece of evidence in support of a hypothesis (change or null) – then Symolac stops seeking evidence and support for the other hypotheses, updates the land cover dataset and reports about the knowledge and reasoning used to update the land cover map.

3.7.   SYMOLAC ver2.0

The expanded Symolac will be able to reason with a greater range of datasets, and will have experts to represent the knowledge through a series of Knowledge Acquisition (KA) exercises. These are illustrated in this section by a "walk through" of a worked example.

3.7.1   *Integration of disparate datasets*

Other sources of data support a wider range of change hypotheses. Examples of these include:

a) Image data of different resolutions can enable the detection of land cover classes;
b) Ordnance Survey maps may be used to determine the presence of a land cover class;
c) land ownership data could indicate the extent of localised management practices;
d) ESA / SSSI maps could similarly indicate local management;
e) Forestry Commission grants and permits for felling and planting from the World Wide Web would indicate where felling and planting events may occur;
f) historical data on the granting of afforestation licenses and payments could indicate the likely timing of felling;
g) Land Capability for Forestry and Agriculture maps indicate the land suitability for commercial conifer forestry and different agricultural practices;
h) historical climate data indicating changes might be used to infer the directions of land cover change, and its existence might be used to more heavily endorse ecological changes that are associated with climate change.

3.7.2   *Knowledge Acquisition and Engineering*

The principal aim was to expand the range of change events about which Symolac can reason and can detect. This resulted in a Knowledge Acquisition (KA) exercise to determine the management and environmental drivers that shape the land cover at any point, the process by which a parcel of land is delineated and ascribed to a particular land cover class in the interpretation. The aim was to identify the sort of information used in the mapping of different land cover classes, to formalise how these various pieces of information are combined in different ways according to the interpreter, the area being mapped, and the tone, texture and contrast within the aerial photograph. These formalisations could then be converted into computer "experts" to embody the knowledge needed to monitor land cover classes, the transition events, or the order with which knowledge is used in determining a land cover type in a particular area.

The KA was performed through two approaches:

1. The AP interpreters were interviewed and asked to describe the process by which they allocated each single feature land cover class.
2. The interpreters were asked to complete two transition matrices at the summary class level (i.e. at the level of "Heather Moor" rather than the single feature classes of "Wet Heather Moor", "Dry Heather Moor" or "Undifferentiated Heather Moor"). One of the transition matrices was to indicate the likely transitions due to recent land use policies and the other due to successional transitions – that is if the management was removed.

The approach was to be as non-specific in the interviews as possible. In an ideal KA design, knowledge from the experts would be derived by example as the knowledge engineer risks biasing the information that expert gives them (Erricsson and Simon, 1984). That is, the experts are asked to complete certain tasks including some surprise tasks, and how they solve the tasks is recorded. In this way the knowledge is acquired without biasing it in a particular direction (towards a particular approach) and the experts impart their knowledge without giving the answers that they think the knowledge engineer wants (Alberdi and Sleeman, 1995). This was very difficult to structure for such a diverse process as AP interpretation and there was limited time available to extract information from each human expert.

The completion of the transition matrices can be seen as an analogy to the Repertory Grid developed by Kelly (1955). This aspect of KA theory uses Personal Construct Psychology (PCP). PCP is a framework for people's knowledge and behaviour based on how we anticipate or predict events using "constructs" (Kelly, 1955). Constructs are created by splitting up experiences into contrasting or dichotomous extremes (Mitchell, 1998). Indicating likelihood of the individual transitions is an example of this.

Analysis of the information provided by the interpreters provided a set of parameters or heuristics that would indicate the presence of any particular land cover type (bio-geographical and management characteristics, API guidelines, including the contrast with other suspected land cover classes, how the different knowledge was differentially prioritised according to the area / land cover type / management etc.) and the likely land cover transitions under contrasting management scenarios. Within this set of knowledge there were areas of human expert agreement and disagreement. The body of consensual knowledge is for the most part that which will be initially used to determine the land cover changes. The information that was indicated by one or two human experts may be used to fine tune the way that evidence is endorsed. For instance some of the human experts have specific spheres of professional interest – a geographic area, a particular land cover type, etc – and in some circumstances their evidence will be more useful in reaching a solution than the consensual opinion.

There were two issues that were implicitly part of the design for the second part of the KA. Firstly that land management is the major driver of land cover change. If the direction and extent of that management could be quantified in some way for a particular parcel of land, then it may be possible to infer management processes over land that is part of the same estate and therefore managed in a similar manner. Secondly, that land ownership in Scotland is extremely concentrated (50% of the area is owned by less than 500 land owners (Wightman, 1995)) and similar trends in management may be expected across large swathes of the semi-natural environment.

The end result is that for each land cover class there is
-   knowledge about the bio-geographical factors that combine to produce that particular land cover;
-   knowledge of the management process by which that cover is modified;
-   knowledge of the how that land cover type appears on aerial photography depending on location, management, biogeography, whatever the position in the classification hierarchy;
-   a description of the heuristic process by which that land cover is mapped.

By describing the characteristics for each possible land cover class and the features of any particular change event, it is possible to organise the knowledge into experts that will co-operate to produce updated land cover datasets.

3.7.3   *A worked example of knowledge application*

The integration of the disparate knowledge and datasets can be illustrated by considering the transitions from one of the LCS88 classes, "Improved Grassland / Pasture". At this point we encounter one of the anomalies of the Land Cover of Scotland dataset: the use of land use descriptors. Improved Grassland implies a managerial system rather than a plant community or species composition. Certainly the species can be inferred, but it is the land use in this instance that was mapped. The Improved Grassland category indicates that the underlying vegetation has been successively improved over many years. This improvement will take the form of fencing it off from the rougher pasture, applying lime, reseeding, and applying fertilisers. There were many grants and subsidies available for this kind of land "improvement" in the post-war period, with the aim of maximising agricultural production. The land cover / land use system is found in between the arable systems and the managed semi-natural systems – it is at the arable fringes. Due to this geographical location, areas that were in fact arable may well have been mapped as Improved Grassland, the interpretation having been biased what the interpreter saw as the overall land use system in operation in that area. The reality of the improved grassland system is that there is an arable component in some fields where silage or turnips may be the crop. The stages in change detection are as follows:
1) Evidence for change at the individual polygon level is generated spectrally. Once there is spectral evidence of a change in the original land cover polygon, a goal is issued to generate change hypotheses.
2) This goal is responded to by the state-and-transition expert, which creates change hypotheses from the potential transitions within the transition matrices.

The transition matrices are at the summary class level (Table 1) and for hypotheses at the individual single feature class level to be generated all the possible changes "to" should be considered at this stage. A subset of the changes at the levels of summary class, general class and single feature class are listed in Table 2.

Table 1 *Summary class transition matrix*

(From)		(To) 1	2	3	4	5	6	7	8	9	10	11	12	13	14	15	16	17	18	19	20	21
Arable	1	1	1	1	1	1	1	0	0	0	1	0	1	1	1	0	0	0	0	0	1	1
Improved grassland	2	1	1	1	1	1	1	0	0	0	1	1	1	1	1	0	0	0	0	0	1	1
Good rough grassland	3	1	1	1	1	1	1	0	0	0	1	1	1	1	1	0	1	0	0	0	1	1
Poor rough grassland	4	1	1	1	1	1	1	1	0	0	1	1	1	1	1	0	1	0	1	0	1	1
Bracken	5	1	1	1	1	1	1	0	0	0	1	1	1	1	1	0	0	0	0	0	1	1
Heather moorland	6	1	1	1	1	1	1	1	1	0	1	1	1	1	1	0	0	0	0	0	1	1
Peatland	7	0	1	0	1	0	1	1	1	0	1	1	1	1	1	1	0	0	0	0	1	0
Montane	8	0	0	0	1	0	1	0	1	0	0	0	0	0	1	0	0	0	0	0	1	0
Rock and Cliffs	9	0	0	0	0	0	0	0	0	1	0	1	1	1	1	0	0	0	0	0	1	0
Commercial Forestry	10	0	0	1	1	1	1	1	0	0	1	1	1	1	1	0	1	0	0	0	1	0
Semi-natural coniferous	11	0	0	0	1	1	1	1	0	1	1	1	1	1	1	0	0	0	0	0	1	1
Mixed woodland	12	1	1	1	1	1	1	0	0	1	1	1	1	1	1	0	0	0	0	0	1	1
Broadleaved	13	1	1	1	1	1	1	0	0	1	1	0	1	1	1	0	0	0	0	0	1	1
Scrub	14	1	1	1	1	1	1	0	1	1	1	1	1	1	1	0	0	0	0	0	1	1
Fresh waters	15	0	0	0	0	0	0	0	0	0	0	0	0	0	0	1	1	0	0	0	0	0
Marshes	16	1	1	1	1	0	1	0	0	0	1	0	1	1	1	1	1	0	0	0	0	0
Salt marshes	17	0	0	0	0	0	0	0	0	0	0	0	0	0	0	0	0	1	0	1	0	0
Dunes	18	0	1	0	1	1	1	0	0	0	1	0	1	1	1	0	0	0	1	0	1	1
Tidal Water	19	0	0	0	0	0	0	0	0	0	0	0	0	0	0	0	0	1	0	1	1	1
Rural development	20	0	1	1	1	1	0	0	0	0	1	0	1	1	1	0	1	0	1	0	1	1
Urban	21	0	1	0	0	0	0	0	0	0	0	1	1	1	0	1	0	1	0	0	1	1

0 indicates change to be unlikely, 1 that a change is possible or likely.

3) Initial support for the change hypotheses is generated. If the precise land cover definition is "improved pasture: no rock outcrops, no clustered farmsteads, no scattered trees" (LCS88 code 90) and the time interval is short (10 years) then it is possible the to endorse the change hypotheses with an initial degree of support.

The short time interval means that the actual transition is unlikely to be to those classes that have scattered trees and rock outcrops. This endorsement can take the form of a degree of belief in the potential change or in the way that order of hypotheses is evaluated – in this case one would expect the degree of scattered trees and rocky outcrops to be the same over a ten year period – and a set of specific change classes, with an initial endorsement of support, is created (see Table 3).

If not enough evidence can be generated in support of the change hypotheses to classes with similar degrees of rocks and trees, then later on in the analysis the goals to look for evidence in support of changes to classes with rocks and trees can be issued.

Table 2 *Expanded transitions from summary classes to general feature classes to single feature classes*

Summary class	General classes	LCS88
Good rough grazing	Smooth grasslands with rushes	150-153
	Smooth grasslands with low scrub	155-158
	Undifferentiated smooth grasslands	160-163
Poor rough grazing	Undifferentiated coarse grasslands	140-143
Bracken	Undifferentiated bracken	170-173
Heather Moorland	Dry heather moorland	110-117
	Wet heather moorland	120-127
	Undifferentiated heather moorland	130-137

→

Single feature classes	LCS88
Dry heather moorland: no rock outcrops, no burning, no scattered trees	110
Dry heather moorland: no rock outcrops, no burning, scattered trees	111
Dry heather moorland: no rock outcrops, burning, no scattered trees	112
Dry heather moorland: no rock outcrops, burning, scattered trees	113
Dry heather moorland: rock outcrops, no burning, no scattered trees	114
Dry heather moorland: rock outcrops, no burning, scattered trees	115
Dry heather moorland: rock outcrops, burning, no scattered trees	116
Dry heather moorland: rock outcrops, burning, scattered trees	117

Table 3 *LCS88 code 90 is likely to change to one of the following*

Original class	Potential change feature classes	LCS88 code
- Improved pasture: - no rock outcrops, - no clustered farmsteads, - no scattered trees"	Smooth grasslands with rushes; no rock outcrops, no scattered trees	150
	Smooth grasslands with low scrub; no rock outcrops, no scattered trees	155
	Undifferentiated smooth grasslands; no rock outcrops, no scattered trees	160
	Undifferentiated coarse grasslands; no rock outcrops, no scattered trees	140
	Undifferentiated bracken; no rock outcrops, no scattered trees	170
	Dry heather moorland; no rock outcrops, no muirburn, no scattered trees	110
	Dry heather moorland; no rock outcrops, muirburn, no scattered trees	112
	Wet heather moorland; no rock outcrops, no muirburn, no scattered trees	120
	Wet heather moorland; no rock outcrops, no muirburn, no scattered trees	122
	Undifferentiated heather moorland; no rock outcrops, no muirburn, no scattered trees	130
	Undifferentiated heather moorland; no rock outcrops, muirburn, no scattered trees	132

4) The next stage is to seek evidence in support of the set of change hypotheses. Each hypothesised change has a set of bio-geographical conditions, AP or management characteristics under which the changes may occur. These can be used to issue goals to gather some evidence of the existence of these conditions with which to support the individual change hypotheses.

In this example a shift to heather moorland would represent a severe shift in management over a ten-year period for Improved Grassland classes and would be unlikely. Consequently the scheduling expert might move the goals for evidence in support shifts to heather moorland hypotheses to the bottom of the list.

5) Having generated this evidence it has to be reasoned with and evaluated

In this example the differentiating factors between the hypothesised land cover types can be shown:

- Wet soil types would indicate a shift to Smooth grasslands with rushes;
- Steeper slopes would indicate a shift to Smooth grasslands with low scrub and to Undifferentiated smooth grasslands;
- Stippled AP texture would indicate a shift to Smooth grasslands with rushes and to Smooth grasslands with low scrub;
- Pale AP tones would indicate a shift to Undifferentiated smooth grasslands.

Contextual information may enable some evidence supporting a change hypothesis to be endorsed with greater weighting. For example, agricultural census returns of the area might be able to provide information about reductions in stock numbers.

If sufficient endorsement for evidence in support of any change hypothesis is generated, and a "good enough" solution can be reached then the process of updating the land cover map can begin. This may involve further image processing and GIS analysis such as the extraction of neighbouring polygons to determine the extent of the change, and will have to be scheduled by the GIS and image processing experts respectively.

6) If a "good enough" solution cannot be reached then Stages 3), 4) and 5) can be iteratively cycled through until either enough evidence is generated in support of a change hypothesis or the extent of the modelled knowledge is reached. Symolac provides explanations of its reasoning process and any changes to the land cover map or where the null hypothesis is accepted due to data or knowledge paucity (despite the evidence of change), are accompanied by textual explanations.

4. Issues And Further Work

Symolac ver 2.0 is gradually being expanded with experts. This in itself is an iterative process of redesigning expert templates so that the generic nature of the application is maintained: image analysis experts are passed parameters by land cover experts, schedulers are passed activation priorities by transition experts, and so on.

At one level the "further work" that could be done to this application is endless – in theory any piece of spatial, managerial or environmental data could be usefully integrated into Symolac. For the purposes of this research the objective is to demonstrate the concepts involved in integrating multiple reasoning mechanisms, methods of evidence endorsement and knowledge of the problem domain into a coherent system that is able to utilise and reason with wide range of disparate datasets and knowledge. This involves the modelling of land cover characteristics, knowledge about the drivers of land cover transitions, the development of the different scheduling and endorsement experts and the refining of the image processing analysis experts.

Ultimately the aim is to have a system that is on-line, is able to constantly monitor land cover change and produce updated land cover datasets.

References

Alberdi, E and Sleeman, D (1995). Taxonomy revision as shift of representational focus. In ECC95 (European Conference on Cognitive Science), pp47-54.

Belward AS, Taylor, JC, Stuttard, MJ, Bignal, E, Mathews, J and Curtis, D (1990). An unsupervised approach to the classification of semi-natural vegetation form Landsat TM data. *International Journal Of Remote Sensing*, 11(3):429-445.

Birnie, RV, (1996). Methodologies for detecting national land cover changes (Scotland), Proceedings of a Technical Workshop, 9-11 July, 1996, Aberdeen, MLURI.

Burrough, P.A. (1986) *Principles of Geographical Information Systems for Land Resources Assessment*. Oxford, Clarendon.

Cherrill, A. and McClean, C. (1995). An investigation of uncertainty in-field habitat mapping and the implications for detecting land-cover change. *Landscape Ecology*, 10(1): 5-21.

Comber, A.J., Birnie, R.V. and Hodgson, M., (1999). Monitoring land cover change in Scotland: the limits of detectability for semi-natural vegetation. *Proceedings of IRLOGI (GIS Ireland 99) Conference*, September 1999, Malahide, pp13-40.

Dry, F.T., Richman, A.G., Hipkin, J.A., Miller, D.R. (1992) *The measurement and analysis of land cover changes in part of the Central Valley of Scotland.* Report to the Scottish Office Environment Department and the Scottish Office Agriculture and Fisheries Department.

Dunn, R, Swanick, C and Harrison, A, (1995). Evaluation of Land Cover of Scotland 1988 Project. Final report to the SOED.

CN, (2000). http://www.nmw.ac.uk/ecn/objectiv.htm, available 13/03/2000.

EEA, (2000). http://www.eea.eu.int/, available 13/03/2000.

Ehlers, M., Greenlee, D., Smith, T. and Star, J. (1991). Integration of remote sensing and GIS: data and data access. *Ph Eng & Remote sensing*. 57(6): 669-675.

Erricsson, K.A. and Simon, H.A., (1984). Protocol Analysis. Cambridge, Mass, MIT Press.

Fuller, R.M., Groom, G.B. and Jones, A.R. (1994). The land-cover map of Great Britain - an automated classification of Landsat thematic mapper data. *Ph. Eng. and RS* 60(5): 553-562.

Gauld, J.H., Bell, J.S., Towers, W. and Miller, D.R. (1991). *The measurement and analysis of land cover change in the Cairngorms*. Report to the Scottish Office Environment Department and the Scottish Office Agriculture and Fisheries Department.

Gensym, (2000). http://www.gensym.com/products/G2.HTM, available 21/08/00.

Goodchild, MF (1994). Integrating GIS and RS for vegetation analysis and modelling: methodological issues. *Journal of Vegetation Science*, 5:615-626.

Green, K., Kempka, D. and Lackey, L. (1994). Using remote sensing to detect and monitor Land-Cover and land-Use change. *Photogrammetric Engineering And Remote Sensing*, 60(3): 331-337.

Hester, AJ, Miller, DR, and Towers, W, (1996). Landscape-scale vegetation change in the Cairngorms 1946-1988: implications for land management. *Biological Conservation* 77:41-51

Holmgren, P and Thuresson, T, (1998). Satellite remote sensing for forestry planning - A review. *Scandinavian Journal Of Forest Research*, 13(1):90-110

Horgan, G, Glaseby, C, Law, A, Skelsey, C, Fuller, R, Elston, D and Inglis, I, (1997). Automating air photo interpretation for the Land Cover of Scotland project, Report for SOAEF, MLURI.

IGBP, (2000). http://www.igbp.kva.se/ , 13/03/2000

Jensen, J.R., (1996). Introductory digital image processing: a remote sensing perspective, Second edition, Englewood Cliffs, Prentice Hall.

Kelly, G.A, (1955). The psychology of personal constructs. New York, Norton.

Lunetta, RS, Congalton, RG, Fenstermaker, LK, Jensen, JR, Mcgwire, KC, and Tinney, LR, (1991). Remote-Sensing And Geographic Information-System Data Integration - Error Sources And Research Issues. *Photogrammetric Engineering And Remote Sensing*, 57(6):677-687

Lyon, JG, Yuan, D, Lunetta, RS and Elvidge, CD (1998). A change detection experiment using vegetation indices. *Photogrammetric Engineering And Remote Sensing*, 64(2):143-150

Macleod RD and Congalton RG (1998).Quantitative comparison of change-detection algorithms for monitoring eelgrass from remotely sensed data. *Photogrammetric Engineering And Remote Sensing*, 64(3):207-216

Mitchell, F, (1998). *An introduction to knowledge acquisition.* AUCS Technical Report, TR9804, University of Aberdeen, Aberdeen.

MLURI, (1993). The Land Cover of Scotland 1988 Final Report. MLURI, Aberdeen.

Salvador, R., Valeriano, J., Pons, X., and DiazDelgado, R., (2000). A semi-automatic methodology to detect fire scars in shrubs and evergreen forests with Landsat MSS time series. *International Journal Of Remote Sensing*, 21(4): 655-671.

Singh A (1989). Digital change detection techniques using remotely-sensed data. *International Journal Of Remote Sensing*, 10(6): 989-1003.

Skelsey, C, (1997). A system for monitoring land cover. Aberdeen University: Thesis, Ph.D. or available by prior arrangement at (http://bamboo.mluri.sari.ac.uk/SYMOLAC/). Contact Chris Skelsey at (chris_skelsey@yahoo.com).

Waldemark, J., (1997). An automated procedure for cluster analysis of multivariate satellite data. *International Journal of Neural Systems*, 8(1): 3-15.

Wightman, A. (1995). Who Owns Scotland? Cannongate, Edinburgh.

Wright, GG, and Morrice, JG, (1997). Landsat TM spectral information to enhance the land cover of Scotland 1988 dataset. *International Journal Of Remote Sensing*, 18(18): 3811-3834

3

Data-Rich Models Of The Urban Environment: RS, GIS And 'Lifestyles'

Richard J. Harris, Paul A. Longley

Formal models can reveal much about the form and function of urban systems. In this respect, a routine ability to measure, share and concatenate new sources of digital data is creating considerable opportunity to develop up-to-date and relevant depictions of urban morphology. In this paper we explore how Ordnance Survey's Code-Point and Address-Point products can be incorporated within a broader framework of RS image classification. Population surface models are created from both Census and postcode data, and, using the unit postcode as a *de facto* standard for data integration, we apply commercial marketing data to obtain models of the residential patterning of the City and County of Bristol, England. Such procedures, we argue, can generate realistic and scientifically grounded models of urban systems, these being a necessary pre-requisite to the effective management of sustainable cities.

1.   Introduction

Formal models of urban systems have the potential to reveal much about the form and functioning of urban settlements (Longley and Harris, 1999). That contention is taken as axiomatic in this paper, although we begin from a starting-point that much of the potential has yet to be realised. The 'era of quantitative geography' from the 1950s until the 1970s was characterised by high levels of analysis and model building. However, the formal structures developed during this period were almost inevitably limited by crude, inappropriate, surrogate data; limited data collection; and coarse zonal aggregations (Longley, 2000). Viewed in the context of the digital data infrastructure that has developed over the last decade, it seems almost foolhardy to suppose such foundations could ever sustain workable, system-wide models. By contrast, our now routine ability to measure, share and concatenate new sources of digital data creates exciting new opportunities to develop relevant and timely depictions of urban morphology, and of urban change.

In this paper we explore the extent to which contemporary digital data sources can be incorporated within a broader framework of 'RS-GIS' integration. We limit our discussion to five sources of data: the 1991 UK Census; a remotely sensed SPOT image; Ordnance Survey's Code-Point™ and AdressPoint™ products; and a 'lifestyles' dataset, acquired from a commercial marketing organisation. Following research undertaken by Mesev (1998) we will place emphasis on the rôle of population surface models (PSMs) as a basis for classifying remotely sensed (RS) images using ancillary sources of socio-economic data.

2.   Socio-Economic Data And The Classification Of Rs Imagery

2.1   The 1991 UK Census: an outline

The confidentiality strictures of the UK Census of Population restrict the dissemination of nearly all Census data to areal aggregations, the smallest of these being the 100,000 or so enumeration districts (EDs) across England and Wales.[1] The design of the EDs for the 1991 Census was principally to facilitate population enumeration through the specific intention of balancing the workloads of enumerators. Urban EDs are characteristically of smaller area but are more densely populated.

Census EDs are administrative units and their design does not allow a direct and precise identification of non-built or non-residential areas at a fine scale. The geography of EDs fits together (tessellates) in such a way that nearly the entire national land surface is allocated to one or another ED. Thus, whilst the principal function of a census is enumeration of the *residential* population (a study of residential spaces), Census EDs will often include *non-residential* land such as inland water bodies, public recreational spaces, derelict land or industrial and commercial premises within their borders.

Moreover, since no measure of variation is given at a sub-areal scale, census-based classifications of the urban environment inevitably imply that population attributes are uniform across each ED (Mesev, 1998). Census geography represents, therefore, a somewhat artificial partitioning of socio-economic space into objects of census administration (see Martin, 1998b). The assumption of internal homogeneity within small areas is questionable since urban populations can be characterised as much by diversity as by uniformity (Mitchell et al., 1998).[2]

---

[1] The exceptions are the Samples of Anonymised Records which give a 1% sample of individual-level Census returns or a 2% sample of household data. However, it is not possible to identify the location of these respondents with any greater precision than the Regional level (e.g. Greater London or Rest of South-East England).

[2] Proposals for the 2001 UK Census include designing Output Areas (OAs) to be more homogenous with respect to categories of tenure (Martin, 1998a). The geography of these OAs will remain a complete tessellation of the land area, however, with no obvious distinction between residential and non-residential spaces being made.

Census data also age and are infrequently up-dated. The next UK Census is scheduled for April 2001 but it seems unlikely that the data will be widely available much before mid-2003 (based on previous experiences, see Openshaw, 1995). By that time the 1991 Census dataset will be over 12 years old and even the 2001 Census information will have dated by two years! The Census offers, therefore, only a crude measurement of fast-changing, urban environments (or other socio-economic systems). We base this paper on the premise that in an age of increased social fragmentation, descriptive models grounded in census or other conventional geodemographic approaches are likely to offer overly limited understandings of human activity patterns and their effects upon the urban environment. We will show how such data can be supplemented with more dis-aggregate and more up-to-date data sources.

2.2   Incorporating socio-economic data within the classification of RS imagery

Remote Sensing (RS) instruments are designed to collect continuous data of physical surface variations, discretized only by the resolution of the aerial scanner (Mesev, 1997; Mesev and Longley, 1999). These data are often refreshed at daily or monthly intervals, particularly when they are generated from orbital observation of the Earth's surface rather than aerial flying. The SPOT image to be classified later in this paper (see Section 4) has a nominal resolution of 12 metres. However, satellite platforms are beginning to obtain data at resolutions of one metre or less. The recently launched IKONOS-2 satellite, for example, incorporates a one metre resolution, panchromatic band (wavelengths 0.45 to 0.90 μm: see Lillesand and Kiefer, 2000). Aerial photography such as the Cities Revealed data (Cities Revealed, 1996) provide digitally rectified images at a ground resolution of one quarter of a metre.

In the view of Mesev (1997), sub-10 metre resolution data will signal the beginning of a new chapter in urban remote sensing. Certainly, RS potentially provides a strong, scientific framework within which to model urban form and to monitor change (Donnay et al., 2000). Nevertheless, it is well known that classified urban land cover does not bear a spectrally identifiable correspondence with urban land use. Bibby and Shepherd (1999 pp. 954-956) suggest land use is defined by a social purpose and not a set of physical quantities. "Unfortunately," they argue, "recourse to mechanical perception as a means of collecting land cover data is not a panacea." It is "by no means the case that inferences can be made about land use from knowledge of physical structure". In practice, therefore, it is necessary to augment RS data with a range of other, ancillary datasets. In summary, RS data are useful for providing outline descriptions of urban form but are less helpful to understanding the functional characteristics of urban settlements. Urban systems evolve from the interplay between function and form (Portugali, 2000), and greater integration of socio-economic and urban RS data should lead to more comprehensive depictions of the urban environment.

As part of the quest to develop an improved interface between RS and GIS techniques, Mesev (1998) proposes a supervised classification strategy linking urban RS land cover data with urban functional characteristics described by the 1991 UK Census of Population. The modelled link is developed from interpolation of census ED data into near-continuous, raster-based surfaces of population density (Martin and Bracken,

1991).[3] Surface representations are often considered to be more realistic representations of population distribution than those given by the unmodified Census geography, and the procedure is particularly effective for dis-aggregating residential from non-residential spaces. Each Population Surface Model (PSM) is, of course, an artefact of the estimation method. Nevertheless, it is likely that the model will provide better estimates of the unmeasured, underlying, population surface density than almost any chloropleth map (Wood et al., 1999).

By interpolating census data into surfaces, Mesev (1998) incorporates the PSM methodology into RS image classification at each of three stages.

- First, prior to classification, when census surfaces are used to assist the selection of class training samples.
- Second, during classification, when census data are used to provide *a priori* probabilities of class membership (e.g low density residential, high density residential) using a Bayesian-modified maximum-likelihood estimator (see Lillesand and Kiefer, 2000 pp.541-544; also Section 4 below).
- Third, post classification, when census surfaces are used to verify the classification of a residential built category from other built land covers.

3.   Population Surfaces: An Outline

Bailey and Gatrell (1995) characterise the population surface model (PSM) algorithm devised by Bracken and Martin (1989) and implemented by Mesev (1998) for RS classification as a form of adaptive kernel estimation. Essentially PSMs are a form of population weighted, kernal density estimation, and the approach is therefore similar to the methodologies described by Silverman (1986) and Brunsdon (1995).

The initial stage of the model assigns the entire population count of the Census ED to a point location – a population-weighted centroid – within the ED. The technique then operates by focusing upon each centroid in turn, calculating the mean inter-centroid distance between each centroid that falls within a search window of radius defined by the user. This mean distance (to which the perimeter of the user-specified search window adapts) is used both as a measure of the unknown space filled by the population within the region and to calibrate a distance decay function. This function gives highest weight to the output grid cells, which are closest to the centroid; hence, least weight to those cells towards the outer edge of the adaptive window. The weights are subsequently used to distribute the population count associated with each centroid across the surrounding region. Each raster cell in the output grid might then receive a population share from one or more centroids at this stage, or remain unpopulated when beyond the area of influence of any of the centroids (Martin, 1996). The redistributed population values give an interpolated measure of population size in each cell. Since the area of each cell is constant so these values are proportional to the estimated population density. The modelling procedure can be adapted to ensure that the population count per

---

[3] Note that it is geographical space, not the population, which is treated as continuous by the software generating these population surface models, and that the output is actually discretized by the size of the raster grid (see Wood et al., 1999).

ED after redistribution is equal to that beforehand (i.e. population totals are constrained within ED boundaries: see Martin, 1996 for detail). However, such a procedure was neither available at the time of generating the population surface estimates available to the academic community at Manchester Information and Associated Services (MIMAS, see http://census.ac.uk/cdu/surpop/), nor for the surfaces used in this paper.

The original rationale for a PSM methodology is described fully in Bracken and Martin (1989) with further detail given by Martin and Bracken (1991). The approach appeals to a physical metaphor whereby each census ED is treated as a body with centre of mass. This centre is deemed the 'high information point'; that is, the greatest mass of population is assumed to be centred around and adjacent to this point (and hence the raster cells nearest the centre always receive the highest allocation of the population under the distance decay function).[4] The limitation of such an assumption is made clear, however, if we consider an old-fashioned set of weighing scales or a child's see-saw, arranged around a central pivot, as illustrated by Figure 1. It is not necessary for the mass to be located at proximity to the fulcrum for the scales to be perfectly balanced. Indeed, the scales are always balanced if Equation 1 is satisfied:

$$M_1 \times d_1 = M_2 \times d_2 \qquad [1]$$

Equation 1 describes the principle of moments. There is absolutely no requirement for either d_1 or d_2 to tend to zero. However, if the PSM interpolation procedure is to occur, then one point does have to be chosen as the one of high information. Treating the tract centroid as the centre of mass – as the high information point – is no doubt preferable to an entirely arbitrary choice of location.

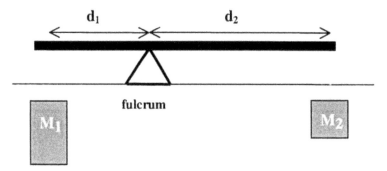

Figure 1. *The Principle of Moments*

---

[4] The surfaces are characterised by a highly positively skewed distribution and usually appear 'spiky' when viewed obliquely. Wood et al. (1999) suggest two methods to overcome this problem, namely local and focal transformations. These methods are not implemented for this paper, however.

4.   Using PSMs For The Classification Of RS Imagery

Mesev (1998) utilises the concept of high information points to select training data for the classification of various satellite images (one SPOT HRV (XS) and two LANDSAT 5 (TM) scenes) showing four medium-size settlements in the UK: Bristol, Swindon, Norwich and Peterborough. The methodology gains credence insofar as the centroids published for the UK purport to locate not the geometric centre of EDs but the population-weighted centre. These points were determined by eye, however, at the Office of Population Censuses and Surveys (now part of the Office of National Statistics) at the time of the census geography design and with a resolution of 10 metres in urban areas (100m in rural areas, Martin, 1996). From visual inspection of the 1,019-centroid records for EDs comprising the City and County of Bristol, overlain upon Cities Revealed, high-resolution aerial photography of the region, it is estimated that 29 (2.8%) are *not* located within the built environment and, therefore, should not be assigned the greatest mass of population (be it persons, households or dwellings). In fact, they should be assigned zero population. There is thus an apparent mismatch between the georeference of some ED centroids and the actual built form of the city. This implies the need for careful user intervention when selecting training data for RS classification from around these point locations. In fact, the entire method of using PSMs to classify RS images can be quite cumbersome requiring at least seven distinct stages to be guided by the analyst:

- generation of the PSM using proprietary software;
- export of the model from the program to RS software (e.g ERDAS Imagine);
- visualisation of the surface within Imagine; geographical rectification of the surface with the RS image to be classified;
- manual selection of the training data from the RS image, the selection to be guided by both the PSM and the locations of the ED centroids;
- class probabilities (e.g. Residential, Non-Residential) to be calculated and specified if required; and
- the visualisation of the final, classified image in Imagine or an alternative, GIS package (the latter option requiring the model be exported between software packages using a file format common to both).

Here we have improved the interface between each stage by developing customised software to proceed from the generation of the PSM to the classification of the RS image more automatically. As we explain below, the program selects, from a SPOT image, all pixels indicated by the PSM to be populated and uses this subset to provide training data for classification. We believe that there is little, if any, justification to discard some of this information on the sole criterion that a pixel is not adjacent to a (subjectively defined) ED centroid. However, we also recognise the concept of high information points and will show how this can be incorporated within the classification procedure. By selecting a much larger number of points for training when compared to the approach used by Mesev (1998), and given that the user no longer fully guides the selection procedure, we risk selecting training data which actually lie outside of built-up areas. Nevertheless, the benefits of a more 'seamless' transition through the classification stages are felt to outweigh such disadvantages. At present this program is written in

FORTRAN, with input and output to ArcView using the GRID format. It would be possible, however, to rewrite both this and the PSM program using either Avenue or Visual Basic script, fully integrating the model generation and classification procedure within the GIS. This is a direction for future research.

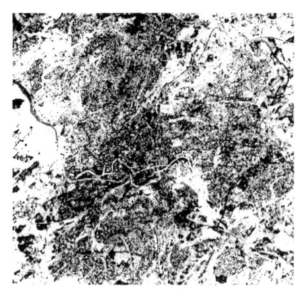

Figure 2. *Unclassified SPOT image showing the City and County of Bristol, England*

Figures 2–9 illustrate each stage of the classification procedure. Figure 2 shows the unclassified, SPOT image for a study region of the City and County of Bristol. The original image is from MIMAS and covers the Bristol, Bath and Frome region of the UK. It is dated March 13, 1993. The scenes held by MIMAS are panchromatic images covering the landmass of the British Isles. The panchromatic mode images data in the spectral range 0.51 to 0.73 µm. The image has a nominal resolution of 12 metres. Figure 3 shows an outline of the PSM generated from the Census count of households per ED with an output grid set to 204 metres (but resampled to 12 metres for compatibility with the SPOT image). The ED centroids or high information points are also illustrated. Having generated the PSM, the next stage is simply a raster overlay operation, selecting as training data for the subsequent RS classification those pixels of the SPOT image which are identified by the PSM to be populated (Figure 4).

The SPOT image is formed by 1,168 columns and 1,084 rows across the Bristol region; a total of 1,266,112 pixels. Each of theses pixels has been ascribed a digital number (DN) in the range 0 to 255, corresponding to the energy detected by the satellite sensor from the original, observed scene (see Lillesand and Kiefer, 2000). Essentially, the panchromatic spectral range (0.51 to 0.73 µm) has been re-scaled over the range 0 to 255.

By using the PSM to select a sub-area of the study region what we are actually obtaining for each class (Residential and Non-Residential) is a subset of all the DN values across the scene (Figure 5). Having obtained each subset, we could now calculate an average DN value per class. We might then assign each of the individual 1,266,112 pixels to the class with mean closest to their own DN values. However, such a procedure discards additional information which may be used for classification; specifically the range of values for each class and a measure of their variance. As a parametric classifier, the maximum likelihood (ML) algorithm computes the statistical

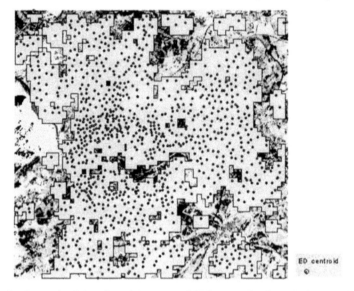

Figure 3. *Stage 2 of the classification: a PSM is overlaid upon the SPOT image*

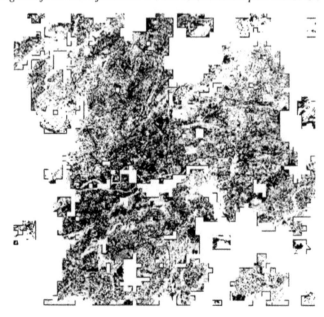

Figure 4. *Stage 3: The PSM is used to select training data*

probability of a given pixel value being a member of a particular land cover class. It is assumed that the training sample may be represented by a Gaussian distribution, which means the probability function is described, completely by the mean vector and the covariance vector (Lillesand and Kiefer, 2000). Although this assumption of Normality is generally reasonable it is not required and not made for the classifications undertaken here. By virtue of using the PSM to select a relatively large sample of pixels from the RS image we can calculate directly the frequency for which each DN value (0 to 255) is associated with a particular class. These frequencies may be expressed as probability values, as in Figure 6, where the sum of the probabilities (the area under each curve) is one. The graphs indicate, for example, a DN of 53 to be more indicative of a Non-Residential area, whilst a DN of 61 is more likely Residential.

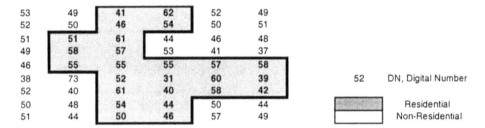

Figure 5.   *Selecting subsets of the DN values*

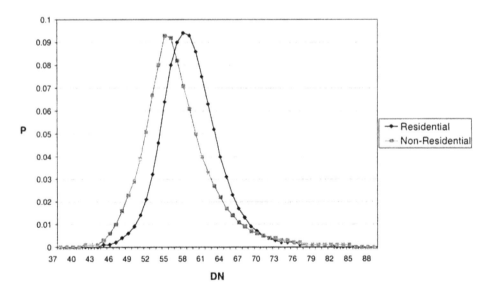

Figure 6.   *Unweighted probability of class membership for different DN values*

The resulting 'spectral signatures', shown in Figure 6, are obtained by giving each pixel within a subset equal weight during the classification procedure. To retain the concept of high information points we can recalculate the probability estimates, weighting the DN of each pixel in accordance with the local population density estimated by the PSM. Assume, for example, that DN 52 and DN 64 are both associated with the class Residential and with equal frequency, although pixels with DN 52 are located more closely to ED centroids on average. Without weighting, the estimated probability of DN 52 representing a Residential area will be equal to that of DN 64. With weighting, DN 52 will have the greater probability. This is because we are weighting by population density, the PSM always interpolating population density to be greatest nearest ED

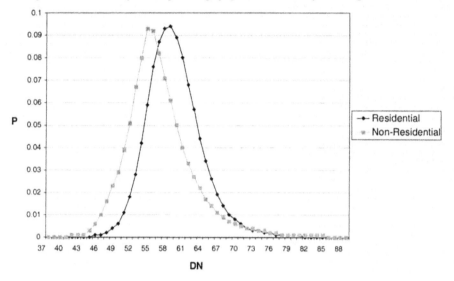

Figure 7.   *Weighted probability of class membership of different DN values*

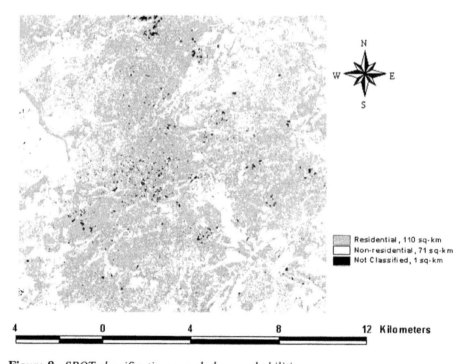

Figure 8.   *SPOT classification, equal class probabilities*

centroids (see Section 3, above). A visual comparison of Figure 7 against Figure 6 suggests the apparent effect of weighting is to increase the separation of the two classes very slightly.

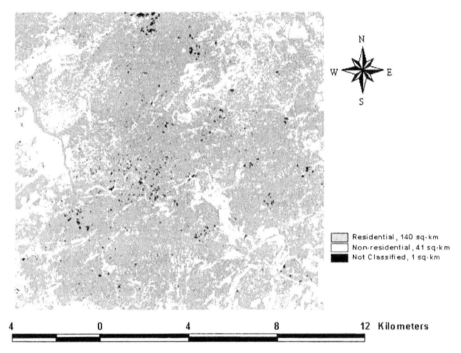

Figure 9. *SPOT classification, unequal class probabilities*

A final stage of classification is to modify the overall class probabilities. Hitherto, each of the classes' probabilities has been equal: the area under each curve (Residential and Non-Residential) is one. In other words, the classification assumes Residential and Non-Residential areas to be equally prevalent within the study region. This is counter-intuitive given that the PSM suggests the area of Residential space to be 115 square-kilometres (km^2), leaving a Non-Residential area of about 67 km^2. The two classes are not equally prevalent, geographically speaking. Assuming the area measures given by the PSM, then the class probability of a pixel being Residential is 0.63 (i.e. 115÷(115+67)) and of being Non-Residential is 0.37 (i.e. 1−0.67).

The class probabilities we have derived are obviously not equal, yet maximum likelihood (ML) classifiers – including the one so far used to classify the SPOT image – typically assume them to be so. Usually, each class would be assigned a probability value of one under an ML classifier, although the correct value is actually 0.5 (since here there are two classes, each of which is as equally likely to occur as the other, and given that the sum of the classes cannot exceed one). Unequal class probabilities can be readily incorporated into the classification procedure on the basis of Bayes law, however (Equation 2, below). Mesev (1998) describes fully the incorporation of these unequal, *a priori* probabilities into the ML classifier. Here, for ease of understanding, we give a simple example which is an extension of and comparable to that given by Campbell (1996 pp.339-341).

Let us say we are interested in obtaining the probability of class Residential (R) given that we have encountered digital number (DN) n. The probability is expressed as $P(R \mid n)$. Similarly, we are interested in obtaining the probability of class Non-Residential (NR) also given n, i.e. $P(NR \mid n)$. To begin with, we assume that the probabilities of obtaining the two classes are equal, $P(R) = P(NR)$, and that we are able to estimate, from our training data, the probability of encountering n given the existence of class R, $P(n \mid R)$; also the probability of n given NR, $P(n \mid NR)$. Bayes law is expressed as follows for our example:

$$P(R \mid n) = \frac{P(R)P(n \mid R)}{P(R)P(n \mid R) + P(NR)P(n \mid NR)} \qquad [2]$$

and;

$$P(NR \mid n) = \frac{P(NR)P(n \mid NR)}{P(NR)P(n \mid NR) + P(R)P(n \mid R)} \qquad [3]$$

Now, since a typical ML classifier assumes each class probability to be one, then $P(R) = 1$ and $P(NR) = 1$, hence:

$$P(R \mid n) = \frac{P(n \mid R)}{P(n \mid R) + P(n \mid NR)} \qquad [4]$$

and;

$$P(NR \mid n) = \frac{P(n \mid NR)}{P(n \mid NR) + P(n \mid R)} \qquad [5]$$

From Figure 7 we estimate for DN 61 that $P(61 \mid R)$ is about 0.09 whilst $P(61 \mid NR)$ is about 0.05. Hence, from Equations 4 and 5, $P(R \mid n) = 0.64$ and $P(NR \mid n) = 0.36$. The ratio of $P(R \mid n) : P(NR \mid n)$ is about 1.8 : 1. Note, that because the denominator in Equation 4 is equal to the denominator of Equation 5 ($a + b$ being equal to $b + a$), so we could have calculated the ratio of $P(R \mid n) : P(NR \mid n)$ from the ratio $P(n \mid R) : P(n \mid NR)$.

As Mesev (1998) notes, the ML classifier is modified by changing the values of $P(R)$ and $P(NR)$, where $P(R) > 0$ and $P(NR) > 0$ and $P(R) + P(NR) = 1$. Unless $P(R)$ and $P(NR)$ remain equal we cannot reduce Equation 2 to Equation 4, or Equation 3 to Equation 5. However, the denominator is still equal in Equations 2 and 4. Therefore, we may say that:

$$\frac{P(R \mid n)}{P(NR \mid n)} = \frac{P(R)P(n \mid R)}{P(NR)P(n \mid NR)} = \frac{P(R)}{P(NR)} \times \frac{P(n \mid R)}{P(n \mid NR)} \qquad [6]$$

Recall that we have estimated, from our surface models, that $P(R) = 0.63$ and $P(NR) = 0.37$; hence the ratio $P(R) : P(NR)$ is approximately 1.7 : 1. From Equation 6 we know that the Residential class will not dominate the classification procedure, provided that the inverse of the ratio $P(n \mid R) : P(n \mid NR)$ is greater than 1.7 : 1 over a reasonable range of DN values. Unfortunately, a close inspection of Figure 7 reveals that the inverse of the ratio $P(n \mid R) : P(n \mid NR)$ is only greater than the ratio $P(R) : P(NR)$ over a limited range, from about 40 to 52. It is unsurprising, therefore, that although the area of Residential space is under-estimated to 110 km^2 when both class probabilities are specified as one (see Figure 8), re-specifying the probabilities over-estimates the Residential area greatly (at 140 square kilometres, see Figure 9).[5] In short, this occurs because the inter-class probabilities vary more than the intra-class probabilities. As a result the class probabilities dominate the classification procedure, and in favour of the Residential class.

The problem we have described seems to point to the heart of the land-cover/land-use dichotomy. The two classes, Residential and Non-Residential, are taken to be mutually exclusive sets: a land area is either Residential or it is not. However, this is only true with respect to the land use classes. We know, from Figures 6 and 7 for example, that with respect to land cover the two classes are not so distinct. They are certainly not mutually exclusive. We would speculate, therefore, that the ratio $P(R) : P(NR)$ (as determined from the PSM) needs to be reduced by some measure of signature separability. Again, this is a direction for future research.[6]

5.   Urban Models At The Postcode Scale

Following an assessment of surface and zonal models of population, Martin (1996 pp. 979, 984) records that a 200m × 200m PSM such as that used by Mesev (1998) and also in this paper "is not an accurate tool for the re-estimation of ED populations." That conclusion was reached by re-aggregating the population counts given to each of the 200 metre grid cells back into groups of EDs. That conclusion was reached by re-aggregating each of the output grid cells back into groups of EDs. After (re-) aggregation, the surface populations ought to tally with the original, unmodified populations (since the surface populations were interpolated from those ED populations in the first instance). However the correlation between the true (unmodified) and replicate (interpolated) population values at the ED scale was actually found to be $r = 0.65$ for a study region in Southampton, England. Furthermore, the outline of residential form provided by the model (see, for example, Figure 3, above) was found to provide a good approximation to

---

[5] This is a 27% increase. In order to validate this result we have compared it against an ISODATA classification undertaken within Erdas Imagine. An initial, unsupervised classification of the image into two classes where equal class probability was assumed gave a 'Residential' area of 67 square-kilometres. Modifying the class probabilities to $P(R) = 0.67$ and $P(NR) = 0.33$ gave a Residential area of 83 square-kilometres. This is a 24% increase.

[6] To some extent Mesev (1998) avoids the problem by first sorting the satellite images into Residential and Non-Residential classes *without* modifying the ML classifier. Having done so, the modification is made to sub-classify the Residential class into four further classes having, in general, more distinct spectral characteristics. The sub-classes are: low density residential, medium density residential, high density residential and tower blocks.

residential geography at a national level only. It compares poorly with residential geography in urban centres where the EDs are typically smaller than the cells of the raster model. Decreasing the size of the raster gird does not lead to a better model. In fact, it achieves the opposite effect, decreasing the correlation value to $r = 0.62$.

An alternative to using the Census counts of households is to use Ordnance Survey's (OS) Code-Point product (originally marketed as Data-Point™). This dataset provides a count of both residential and commercial mail delivery points per unit postcode. The number of residential delivery points (RDPs) is, in general terms, the number of letterboxes into which private mail can be received. On average, there is a mean of 16 delivery points per postcode, with the median being 12 (Shepherd and Ming, 1998). Individual households might share a letterbox, particularly within houses in multiple occupation (HMOs). Consequently, the number of delivery points is not always a direct measure of the true number of households or of population density. Nevertheless, Code-Point can be easily modelled by the PSM program since a population-weighted centroid is given for each postcode (derived from OS' Address-Point data: see Section 6 below). Indeed, we anticipate Code-Point to offer a number of advantages over the Census for the following reasons (Harris and Longley, 2000):

- Code-Point has an apparently finer resolution. The majority of the centroids are precise to one metre (OS, 1997) as opposed to the ED centroids which are subjectively defined and which have a 200 metre resolution. Preliminary analysis has suggested PSMs may be generated from Code-Point information with an output grid size of 50 metres (here we shall use 48 metres, resampled to 12 metres).

- Royal Mail is interested only in premises that receive mail. Delivery points do not include, therefore, open water or derelict land. Furthermore, industrial premises can be distinguished from residential properties.

- The unit postcode is a spatial referencing system common to many other social-economic datasets (including the lifestyles dataset to be considered in Section 6, below). Code-Point offers, therefore, a framework for integrating different datasets whilst avoiding some of the privacy and confidentiality implications of individual or household-level analysis (see also Raper et al., 1992).

Figure 10 shows the outline of a PSM generated using Code-Point and a grid size of 48 metres, overlain upon the outline of the PSM generated using 1991 Census data with a grid size of 204 metres (and also shown in Figure 3). The Code-Point model estimates the Residential area of Bristol to be 80 km^2 against a Non-Residential area of 99 km^2. Under the Census model it is the Residential area, which is larger, at 114 square-kilometres. The Non-Residential area is 65 square-kilometres. The comparison suggests that it is the postcode-based PSM which gives the more precise outline of residential geography, and this is confirmed when the outline of the postcode-based PSM is overlain upon the Cities Revealed aerial photography for a part of the study region. It is evident that the postcode geography is better able to delimit the Built from the Not Built and also Residential from Commercial (under the Census-based PSM the full extent of this area is modelled as Residential).

It is somewhat surprising therefore that there is very little change in the classification output given equal class probabilities once the less aggregate Code-Point information is substituted as the input to the PSM (to select training data) instead of the

Census count of households. Since the postcode-based PSM has a resolution approximately four times greater than the Census-based PSM, so the classification methodology appears less sensitive to the scale changes than we originally anticipated. However, in the bottom-part of Figure 11 we have visually exaggerated the areas where there are differences, to emphasise the spatial patterning. It is evident that may of the pixels classified as Residential, but under the Census-based classification *only*, are located within Bristol's Central Business District – areas where residential (mail) delivery points are either rare or do not occur at all.

Finally, we know (from Figure 10) that the two PSMs vary in the amount of space they fill as Residential. If these areal measures are used to assign unequal class probabilities then the resulting classifications vary more greatly than before. In Figure 12 the Residential area is now shown to be 140 square-kilometres under the Census based classification, but remains at 109 square-kilometres under the postcode-based classification.

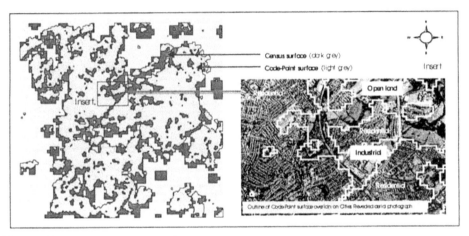

Figure 10. *The Census and Code-Point models compared (source: Harris and Longley, 2000)*

6. Modelling Urban Lifestyles

Ordnance Survey's Address-Point product gives a national grid reference for each mail delivery point in the UK. A precision of 0.1 metres is reported for the majority of cases (OS, 1996) and it is, in fact, the spatial average per postcode of these points which provides the centroid given in the Code-Point set. Following work undertaken by Longley and Mesev (2000), two files based on Address-Point were available to us: the first containing only commercial delivery points; the second only non-commercial (i.e. RDPs). Figure 13 shows the non-commercial delivery points within the Bristol study region, overlaid upon the classified SPOT image as described in the preceding Section and illustrated in Figure 12.[7]

---

[7] i.e. Residential/Non-Residential classification, based on PSM-Code-Point combination and unequal class probabilities.

Using the Address-Point information we have undertaken a basis validation of the SPOT classification. We know, for example, that 74% of the non-commercial delivery points given by Address-Point are also located within areas labelled as Residential under our classification. This implies an omission error of 26%. The cause of this error is likely to

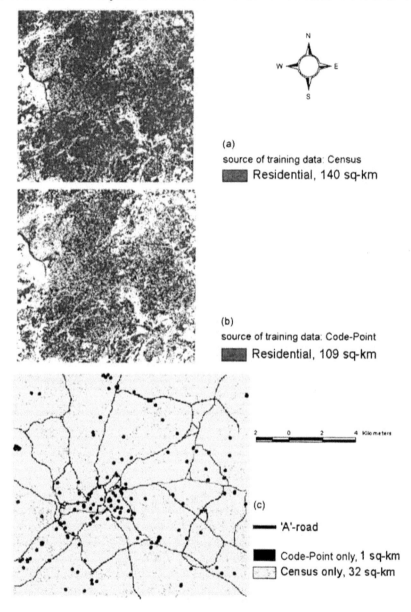

(a)
source of training data: Census
Residential, 140 sq-km

(b)
source of training data: Code-Point
Residential, 109 sq-km

(c)
'A'-road

Code-Point only, 1 sq-km
Census only, 32 sq-km

Figure 11. *The Census and postcode classifications compared; equal class probabilities. (a) Census based classification of SPOT image; (b) Code-Point based classification of SPOT image; (c) Areas classified as Residential under one but not both classifications*

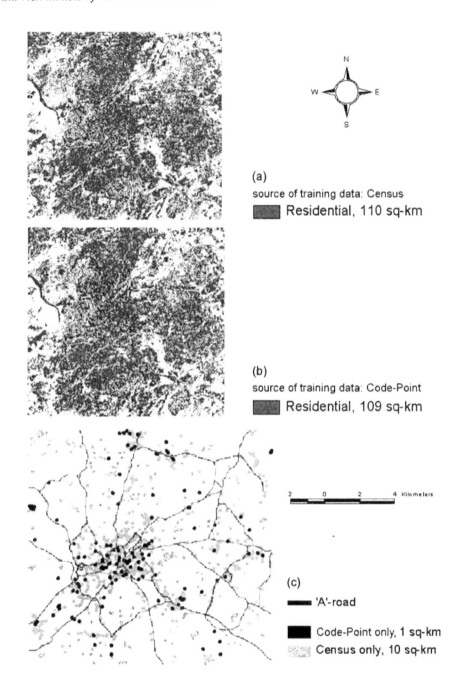

(a)
source of training data: Census
Residential, 110 sq-km

(b)
source of training data: Code-Point
Residential, 109 sq-km

(c)
'A'-road

Code-Point only, 1 sq-km
Census only, 10 sq-km

Figure 12. *The Census and postcode classifications compared; unequal class probabilities. (a) Census based classification of SPOT image; (b) Code-Point based classification of SPOT image; (c) Areas classified as Residential under one but not both classifications*

be primarily due to the vagaries of classification, although other possible causes arise from the georectification of the SPOT image to National Grid co-ordinates and, similarly, during the georeferencing of the mail delivery points for Address-Point. With regards an error of commission, it is evident that our classification over-estimates the area of residential space within Bristol by mis-classifying the Avon Gorge and River amongst other features. It is also more a sifting of the Built from the Not Built as opposed to Residential from Commercial. If a 48 metre buffer zone is generated around the non-commercial delivery points given by Address-Point, then the area within the buffer, which we may estimate to be Residential, is found to be of 97 square-kilometres. The area of residential space estimated from the RS classification is greater, at 109 square-kilometres. On this basis the error of commission is about 12% (i.e. (109−97)÷97), although the size of the apparent error is in part dependent on the specification of the buffer zone.

Figure 14 illustrates a classification of the SPOT image where pixels falling beyond the buffer zone are treated as 'Background' or 'No Data' values. The remaining values are classified into two classes: 'Residential Terraced properties or Flats'; and 'Other Residential'. The classification procedure is similar to that described in Section 5, except we have now introduced a new source of data and generated a further PSM to select training data for the Terraced or Flats class. These data are from a so-called 'lifestyles' database, compiled by a commercial marketing firm on the basis of information received in response to a national, postal consumer survey (see Sleight, 1998). The respondents to such surveys form, in the main, a self-selecting sample of the population. Yet, whilst it is true to say that the collection and dissemination of these data 'flies-in-the-face' of many scientific orthodoxies (Longley, 1998; Goodchild and Longley, 1999), still the data are disaggregate, detailed and up-to-date (most are up-dated annually). The data analysed here are described fully elsewhere (Longley and Harris, 1999; Harris, 1999a; Harris, 1999b) so we describe them only briefly, in Table 1. We argue that incorporating these data into a broader framework of urban environmental modelling brings three related benefits:

- first, the approach is illustrative of how new sources of purportedly 'unscientific' data may be grounded within a more rigorous and orthodox data infrastructure;
- secondly, and reciprocally, data-rich models might be obtained that are more sensitive to context than their census-based forbears but which do not sacrifice generality;
- thirdly, and of particular relevance to the context of this paper, the ancillary data provided by lifestyles might lead to improved classifications of RS imagery.

Figure 15 shows a thematic multiplication of the preceding SPOT classification by the Address-Point coverage. We have, in effect, sorted high density and multiple occupancy housing (i.e Terraced or Flats) from other housing-types. The result is to classify 33% of the residential delivery points as Terraced or Flats, leaving the remaining 67% as Other Residential. The 1991 Census indicates 135,828 dwellings (64%) in the study region to be either Terraced or Flats whilst 76,436 dwellings (36%) are either Detached or Semi-detached. These values imply that nearly two thirds of dwellings in Bristol (either Terraced or Flats) occupy one third of the residential space. In other words, the density is twice that of Detached or Semi-detached properties.

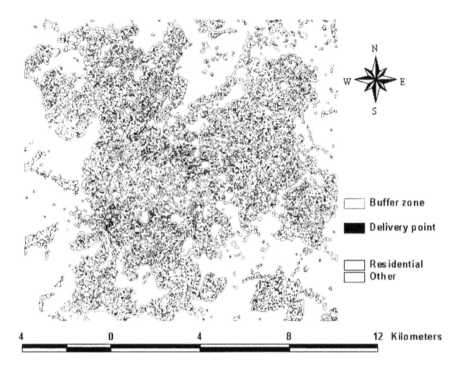

Figure 13. *Residential delivery points overlaid upon SPOT classification*

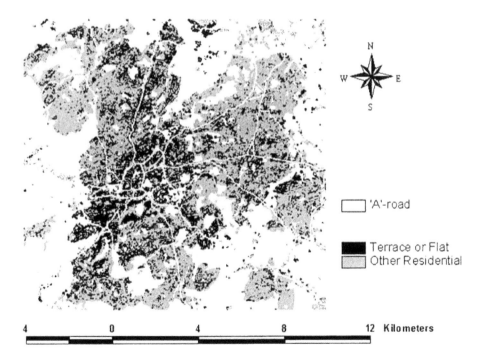

Figure 14. *Classification of SPOT image into two Residential classes*

	Lifestyles data	Census data
Date of survey	Summer 1996	April 1991
Unit of aggregation	Household but georeferenced to postcode	Census ED
Population coverage	16% of households	100% or 10%, depending on variable
Sampling	Self-selecting	100% or 10% random sample
Information types	Wide range of socio-economic, behavioural and consumption indicators	Socio-economic and demographic
Compatibility with postcode geography	Perfect	70-80% (using ED-to-postcode directories)
Bias	Under-enumeration of HMOs and non-ER registrants. Response bias likely to be multivariate and difficult to quantify.	The 'missing millions'

ED, enumeration district; HMOSs, houses in multiple occupation; ER, Electoral Register

Table 1. *Some characteristics of the lifestyles dataset compared against the 1991 Census (based on Longley and Harris, 1999)*

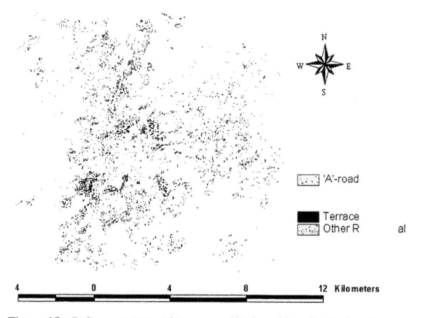

Figure 15. *Delivery points within areas of high and low density housing*

7.   Conclusion: Towards The Model City

The UK Urban Task Force's report to Her Majesty's Government (Urban Task Force, 1999) speaks of a post-industrial age where powerful drivers are at work, changing our towns and cities beyond recognition. They identify one main factor as social transformation: changing life patterns reflecting increasing life expectancy and the development of new lifestyle choices. New trends, in particular the increase in the number of one-parent households, point to a diverse and growing consumer group. Indeed, the report argues for an urban environment that must foster and protect the diversity of its inhabitants whilst ensuring all enjoy access to the range of services and activities that constitute the best of urban life. At the heart, there is a vision for a culturally diverse and socially equitable city with a commitment to positive community relations and ethnic diversity. It is understood, however, that the certain communities remain excluded and face serious issues of marginalisation.

Processes of fragmentation and diversification within the social areas of cities have, we believe, made the 'mosaic metaphor' of census-based, small area analysis increasingly untenable (see Johnston, 1999). However, it is also our opinion that technology does not just cause consumption to fragment but also empowers us to provide ever-richer depictions of the diversity of population characteristics and behaviour within city systems, and also to identify other areas where forces of 'ghetto-isation' appear to prevail. Here we have explored how a number of commercially available datasets can be brought together within a broader framework of integrating RS and GIS techniques. A model of the residential form of the City and County and Bristol has been created. This model might be extended to incorporate LIDAR (light detection and ranging) data of terrain elevation. The potential to generating functional understandings of urban settlements from the lifestyles data remains to be fully exploited.

Data integration raises the problems of identifying sources of error within each dataset and of quantifying the effects of error propagation upon the composite data. This is not easily done, although the microsimulation approach demonstrated by Birkin and Clarke (1996) shows how the vagaries of lifestyle data collection may be assessed against a framework of more orthodox social surveys. Meta-data standards may also help to resolve these problems. Nevertheless, undertaking analysis at a personal level certainly raises the thorny issues of individual confidentiality and privacy (Curry, 1998; 1999). These sit uncomfortably with the European Directive EU 95/46/EC which, having effect in law in all member states, requires all 'personal data [digital information about people] must be relevant, adequate and not excessive; and shall be up to date' (Crayton, 1998). Individuals have the right in principle, although less so in practice, not to be subject to an administrative or private decision involving his or her conduct and which has, as its sole base, the automatic processing of personal data defining his or her profile or personality. Whereas Census ED is too coarse a resolution for urban analysis of the local environment, it ought to be possible to establish a zoning scheme which balances the need of the urban planner with the privacy of the individual. In practice, postcodes at the unit level seem to have become the *de facto* standard in the UK (Raper et al., 1992). Proposals for a more 'intelligent' design to the 2001 UK Census output areas offer a much improved match between Census and postal geographies than is available at present (see Martin, 1998).

To conclude, we follow Rizzi (1999) by arguing that the inadequacies of self-organisation in cities necessitates some kind of intervention, and, in order to intervene, some knowledge of city dynamics is required. Although the models we have presented are temporally contingent 'snapshots' of Bristol, the data they are based on are regularly implemented. As such, these datasets provide opportunity for a longitudinal approach to be adopted. Batty (1999) argues that it is the form of cities that must be manipulated in the grand quest to improve urban conditions. A necessary pre-requisite to this manipulation is to establish realistic and scientifically grounded models of urban morphology. Our on-going investigations have revealed how new, relevant, and timely lifestyles data may be 'tied' to other framework data such as those provided by Remote Sensing or Ordnance Survey's Address-Point. We believe that such approaches offer the prospect of creating vastly enhanced models of the form and functioning of systems which can be implemented into the management of 'sustainable cities' (Haughton and Hunter, 1999; Sattherwaite, 1999).

Acknowledgements

The research was funded under the (UK) Natural Environment Research Council, Urban Regeneration   and   the   Environment   programme (http://www.nerc.ac.uk/es/urgent.htm), grant GST/02/2241. The authors are grateful to Ordnance Survey (GB) and to a second commercial data vendor for the supply of data used in the analysis here.

References

Bailey, T. C. and Gatrell, A. C. (1995) *Interactive Spatial Data Analysis*, Harlow: Longman.
Batty, M. (1999) Editorial, *Environment and Planning B* **26**, 475–476.
Bibby, P and Shepherd, J. (1999) Monitoring land cover and land-use for urban and regional planning, in P. A. Longley, M. F. Goodchild, D. J. Maguire and D. W. Rhind (Eds), *Geographical Information Systems: principles, techniques, management, and applications* (2nd edn.), 953–965, Chichester: Wiley.
Birkin, M. and Clarke, G. (1996) Using microsimulation to synthesize census data, in S. Openshaw (Ed.), *Census Users' Handbook*, 363–387, Cambridge: GeoInformation International.
Bracken, I. and D. Martin (1989) The generation of spatial population distributions from census centroid data, *Environment and Planning A* **21**, 537–543.
Brunsdon, C. (1995) Estimating probability surfaces for geographical points data: an adaptive kernal algorithm, *Computers and Geosciences*, **21**, 877–894
Campbell, J. B. (1996) *Introduction to Remote Sensing* (2nd edn.), London: Taylor and Francis.
Cities Revealed (1996) Orthorectified, aerial photography. http://www.crworld.co.uk
Crayton, T. (1998) Protect yourself, *New Perspectives* **18**, 24–27.
Curry, M. R. (1998) *Digital Places: living with geographic information technologies*. London: Routledge.
Curry, M. R. (1999) Rethinking privacy in a geocoded world, in P. A. Longley, M. F. Goodchild, D. J. Maguire and D. W. Rhind (Eds), *Geographical Information Systems: principles, techniques, management, and applications* (2nd edn.), 757–766, Chichester: Wiley.
Donnay, J. P., Barnsley, M. J. and Longley, P. A. (2000, Eds.) *Remote Sensing and Urban Analysis*. London: Taylor and Francis.

Goodchild, M. F. and Longley, P. A. (1999) The future of GIS and spatial analysis, in P. A. Longley, M. F. Goodchild, D. J. Maguire and D. W. Rhind (Eds), *Geographical Information Systems: principles, techniques, management, and applications* (2nd edn.), 567–580, Chichester: Wiley.

Harris, R. (1999) Geodemographics and geolifestyles: a comparative review, *Journal of Targeting, Measurement and Analysis for Marketing*, **8**, 164-178.

Harris, R. (1999) Modelling lifestyles in an urban system: using lifestyle data for geographical research, in P. Rizzi (Ed.), *CUPUM '99: computers in urban planning and urban management, on the edge of the millenium*, 99, Milan: FrancoAngeli.

Harris, R. J. and Longley, P. A. (2000) New data and approaches for urban analysis: modelling residential densities, *Transactions in GIS*.

Haughton, G. and Hunter, C. (1999) *Sustainable Cities*, London: the Stationary Office Books.

Johnston, R. J. (1999) Geography and GIS, in P. A. Longley, M. F. Goodchild, D. J. Maguire and D. W. Rhind (Eds), *Geographical Information Systems: principles, techniques, management, and applications* (2nd edn.), 39–47, Chichester: Wiley.

Lillesand, T. M. and Kiefer, R. W. (2000) *Remote Sensing and Image Interpretation* (4th edn.), Chichester: Wiley.

Longley, P. (2000) Spatial analysis in the new millennium, *Annals of the Association of American Geographers* **90**, 157–165.

Longley, P. A. (1998) Foundations, in P. A. Longley, S. M. Brooks, R. McDonnell and B. Macmillan (Eds), *Geocomputation: a primer*, 3–15, Chichester: Wiley.

Longley, P. A. and Harris, R. J. (1999) Towards a new digital data infrastructure for urban analysis and modelling, *Environment and Planning B* **26**, 855–878.

Longley, P. A. and Mesev, V. (2000) On the measurement and generalisation of urban form, *Environment and Planning A* **32**, 473–488.

Martin, D. (1996) An assessment of surface and zonal models of population, *International Journal of Geographical Information Systems* **10**, 973–989.

Martin, D. (1998) 2001 Census output areas: from concept to prototype, *Population Trends* **94**, 19–24.

Martin, D. (1998) Automatic neighbourhood identification from population surfaces, *Computers Environment and Urban Systems* **22**, 107–120.

Martin, D. and Bracken, I. (1991) Techniques for modelling population-related raster databases, *Environment and Planning A* **23**, 1065–1079.

Mesev, V. (1997) Remote sensing or urban systems: hierarchical integration with GIS, *Computers Environment and Urban Systems* **21**, 175–187.

Mesev, V. (1998) The use of census data in urban image classification, Photogrammetric Engineering and Remote Sensing 64, 431–438.

Mesev, V. and Longley, P. A. (1999) The rôle of classified imagery in urban spatial analysis, in P. M. Atkinson and N. J. Tate (Eds), *Advances in Remote Sensing and GIS* Analysis, 185–206, Chichester: Wiley.

Mitchell, R, Martin D. and Foody, G. (1998) Unmixing aggregate data: estimating the social composition of enumeration districts, *Environment and Planning A* **30**, 1929–1941.

Openshaw, S. (1995, Ed.) *Census Users' Handbook*. Cambridge: GeoInformation International.

Ordnance Survey (1996) Address-Point sample data user guide. Southampton: OS.

Ordnance Survey (1997) Code-Point sample data user guide. Southampton: OS.

Portugali, J. (2000) *Self-oganization and the city.* Berlin: Springer.

Raper, J., Rhind, D. and J. Shepherd (1992) *Postcodes: the new geography,* Harlow: Longman.

Rizzi, P. (1999) Computers in urban planning and urban management at the turn of the millenium, *Computers Environment and Urban Systems* **23**, 147–150.

Sattherwaite, D. (1999) *The Earthscan Reader in Sustainable Cities,* London: Earthscan Publications.

Shepherd, J. and D. Ming (1998) *Postcodes into Geography,* London: Geographic Information Services.

Silverman, B. W. (1986) *Density Estimation for Statistics and Data Analysis,* London: Chapman and Hall.

Sleight, P. (1997) *Targeting Customers: how to use geodemographic and lifestyle data in your business* (2nd edn.), Henley-on-Thames: NTC Publications Ltd.

Urban Task Force (1999) *Towards an Urban Renaissance: final report of the urban task force,* London: E and FN Spon.

Wood, J. D., Fisher, P. F., Dykes, J. A., Unwin, D. J., Stynes, K. (1999) The use of the landscape metaphor in understanding population data, *Environment and Planning B* **26**, 281–295.

4

Extraction of Field Boundary Information: Using Satellite Images Classified by Artificial Neural Networks

Taskin Kavzoglu, Jasmee Jaafar and Paul M. Mather

Remotely sensed images and the thematic maps derived from such images are invaluable sources of information for GIS databases, in terms of providing spatial and temporal information about the nature of Earth surface materials and objects. One of the techniques that has emerged recently, and which has made a great impact on scientific community, is that of artificial neural networks (ANNs). ANNs have been found to be more robust than conventional statistical methods. ANNs have the advantage of being employed in almost all the stages of a GIS system, such as the data preparation, analysis and modelling stages. This study describes a method to extract accurate field boundary information from thematic maps produced from ANN classification results. A feed-forward network structure that learns the characteristics of the training data through the backpropagation learning algorithm is employed to classify six land cover features present within the scene. This study also illustrates the role of ANNs in classifying land cover objects. A number of factors affecting classification accuracy, including the determination of the optimum network structure, are discussed. It is observed that classification accuracy of up to 90% is achievable for thematic maps produced by ANNs.

1. Introduction

As well as being primary sources for regional and global scale GIS applications, remotely sensed images and the thematic maps, extracted from such images provide top-level information for the inventory, monitoring, and management of natural resources. Given the diversity and heterogeneity of the natural and human-altered landscape, it is obvious that the time-honoured and laborious method of ground inventory is inappropriate for mapping land use and land cover over large areas (Civco, 1993). Therefore, the use of remotely sensed images is essential to form a GIS database for regional or global scale studies.

In particular, thematic maps of land cover, produced by automatic classification systems, are of considerable importance in many environmental GIS applications. A promising approach to the identification of land cover types is the delineation of the boundaries of land parcels. A number of approaches to the problem of automatic boundary detection using line following methods are reported in the literature (For example, Fay and Miller, 1991; Gruen and Li, 1995; Merlet and Zerubia, 1996). Such an approach can be used to extract field boundary information. Even though limited success is reported, due to the complexity of the structure of remotely sensed images, the challenge to develop automated line detection algorithms still remains. On the other hand, statistical techniques have been widely used for classifying land cover objects shown on remotely sensed images. The demands imposed by the use of multispectral, multitemporal, multisensor and ancillary data, such as elevation, slope and aspect information, have turned attention towards the use of new and robust classification techniques, one of which is the artificial neural network (ANN).

Since GIS systems generally involve very large amounts of spatial data, derived from different sources, including remote sensing, topographic, hydrographic and surveying, the accurate analysis of such data is essential for the decision making process. Artificial neural networks have been used intensively in many fields for analysing large quantities of spatial data, which frequently contain noise and information that is not relevant for the purpose of the study. Because of their powerful characteristics, ANNs exhibit a great potential for the analysis of such data, especially for interpolation and extrapolation tasks. Currently, the use of ANNs in GIS is limited, yet there exist many possible application areas. They can be, for instance, used for natural hazard prediction, land suitability assessment and spatial interpolation problems. Werschlein and Weibel (1994), for example, investigate the use of neural networks for line generalisation problem, and conclude that neural networks provide more holistic solutions to line generalisation, avoiding the need to split the generalisation process into discrete operators, such as simplification, smoothing and enhancement. As artificial neural network methods generally provide high level of classification accuracy, they can be employed to perform cumbersome analysis and modelling tasks. Wang (1994) used a three-layer feed forward neural network to estimate absolute and relative land suitability rates for fields in conjunction with an ARC/INFO GIS. He concluded that neural networks are effective for agricultural land suitability assessment in a GIS context (about 84% of overall accuracy was achieved).

So far, little attention has been given to the possibility of retrieving polygonal boundary information by vectorising the thematic maps produced by ANNs. Such vector maps could be useful to support other mapping applications, such as the construction of Digital Surface Models by incorporating height information acquired by other sensors (Jaafar *et al.*, 1999), and accurate crop yield estimation. Furthermore, the accuracy of an ANN classification can also be assessed qualitatively by comparing the produced vector map with a manually digitised land cover vector map.

1.1.   Test Site And Data

A small test site (Figure 1), with an area of 3.73 km², was selected for use in this study. The test site is located near the town of Littleport, in Eastern England, in a rich agricultural area. The study area comprises a number of farmlands of varying crops that also includes man-made objects, such as buildings and roads. Multisensor and multitemporal data, including two

Landsat TM and four SPOT HRV images, are used to classify main crops; wheat, potato, sugar beet, onion, peas, and daffodils.

Ground truth was collated from Field Data Printouts collected from individual farmers and their representative agencies. The images were registered to the Ordnance Survey of Great Britain's National Grid using the ERDAS Imagine image processing software (version 8.3) by applying a first-order polynomial (linear) transformation. The RMSE (Root Mean Square Error) values estimated for image transformations were less than one pixel. In the resampling process of the co-registration stage, all images are reduced to 30 metre spatial resolution.

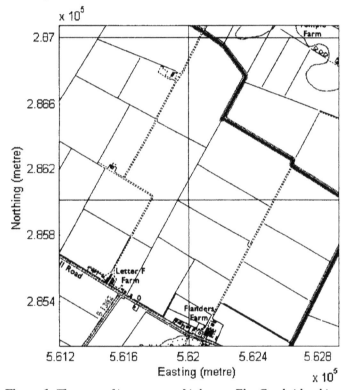

Figure 1. *The area of interest near Littleport, Ely, Cambridgeshire.*

Coordinates refer to OSGB National Grid (reproduced from Ordnance Survey mapping with the permission of The Controller of Her Majesty's Stationery Office, Crown Copyright, ED 273554).

Training and test data were generated from a 285 by 285 pixel image, using rectangular areas in order to include the variation of each crop in the fields. At least three samples (i.e three separate blocks of pixels) were taken for each land cover class to form the training data files. The training pattern file included 2262 pixels, whilst the test pattern file comprised 2204 pixels.

1.2. Artificial Neural Networks

Artificial Neural Networks (ANNs) have been successfully used in many fields for a wide variety of applications. ANNs are heuristic algorithms in that they can learn from experience via samples and are subsequently applied to recognise new, unseen, data. These systems are intended, in an extremely simple way, to imitate the behaviour of the network of neurons in the human brain. The reasons for their popularity rest on their unique advantages, including their non-parametric nature, arbitrary decision boundary capabilities, ability to handle noisy data, and easy adaptation to different types of data including soil, vegetation and geology maps. The power of the network depends on how well it generalises to new data following training. This is the main criterion for judging the performance of a network. The major factors affecting the generalisation capabilities of ANNs are the size of the training data, training time, and the structure of the network. Artificial neural networks learn the characteristics of the training data typically in an iterative way, and are, thus, data-dependent. The characteristics of the training data, therefore, have a considerable influence on the accuracy of the subsequent classification.

The basic element of the ANN is the processing node, which corresponds to the neuron of the human brain. Each processing node receives and sums a set of input values, and passes this sum through an activation function which provides the output value of the node, which in turn forms one of the inputs to a processing node in the next layer of the ANN. Whilst activation functions, also known as transfer functions, are employed to decrease the number of iterations, they introduce non-linearity into the network and thus improve its performance. Although there are several activation functions suggested in the literature, a sigmoid function is generally preferred (Kavzoglu and Mather, 1999).

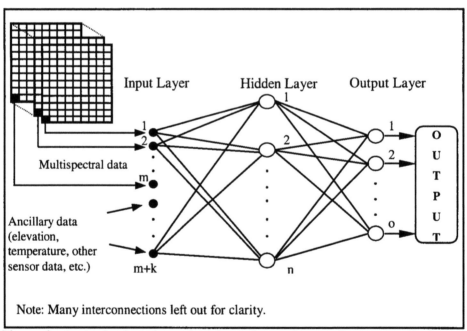

Figure 2. *A simple three layer feed-forward neural network structure* (Paola and Schowengerdt, 1995).

Processing nodes make up a set of fully interconnected layers, except that there are no interconnections between nodes within the same layer in the standard feed-forward backpropagation neural networks. The structure of a feed-forward artificial neural network includes three types of layers: input layer, output layer and hidden layer (Figure

2). The input layer introduces the distribution of the data for each class to the network. The output layer is the final processing layer that has a set of values (or codes) to represent the classes to be recognised. The layers between the input and output layer are called hidden layers. These hidden layers, of which there may be only one, perform the basic calculations. Through these layers the internal representations of the input patterns can be produced. A typical neural network consists of one input layer, one or two hidden layers and one output layer.

Each neuron in the input layer represents one of the input features, such as a SPOT HRV band, while each neuron in the final layer corresponds to one of the output classes. All inter-node connections have associated weights, which are usually initially randomised. When a value passes through an inter-connection, it is multiplied by the weight associated with that inter-connection. The weights in the network determine class boundaries in the feature space. However, there is some evidence that the initial values of the weights may influence the final classification accuracy significantly (Ardö *et al.*, 1997 and Skidmore *et al.*, 1997).

The back propagation algorithm, also called the generalised delta rule, is an iterative, gradient descent training procedure. It is carried out in two stages. In the first stage, after initialising all the network weights randomly, the input data are presented to the network and propagated forward to estimate the output value for each pattern set. In the second stage, the difference (error) between known and estimated output is fed backward through the network and the weights are changed in such a way that these differences are minimised. The whole process is repeated iteratively with new weights until the error is minimal or lower than a given threshold value.

1.3. Determining The Optimum Network Structure

Three major approaches have been proposed to determine the optimum network structure with the aim of achieving high accuracy results. The first approach starts with a small network and new neurons are added to the hidden layer of the network until the error is reduced to a certain level set by the analyst. The techniques using this approach are called constructive techniques, of which the most popular is the cascade correlation algorithm proposed by Fahlman and Lebiere (1990). Such methods are time demanding because a number of networks must be trained for comparative purposes, while small networks are more sensitive to initial conditions and the learning parameters.

The second approach begins with a reasonably large network and reduces its size by deleting a hidden node, or an interconnection between the nodes, until satisfactory learning occurs. Such algorithms are called pruning algorithms, which have the advantage that relatively large networks learn quickly and are less sensitive to initial conditions and learning parameters than are small networks. Most popular pruning techniques used in the literature are known as magnitude based pruning, optimum brain damage (Le Cun *et al.*, 1990) and optimum brain surgeon (Hassibi and Stork, 1993).

There also exist some techniques that employ both constructive and pruning strategies. These techniques couple the pruning and constructive techniques in a way that the size of a small network is increased during training until a reasonable solution is reached and then a counter strategy is applied to reduce the size of the network employing pruning methods. Thus, a smaller and faster network that has, at the same time, higher generalisation capabilities, is produced. Such an algorithm is proposed by Hirose *et al.* (1991). This procedure is employed in this study to determine the optimum network structure. More information about the issues related to the determination of optimum network structure can be found in Kavzoglu (1999).

1.4.   Selection Of Optimum Subsets

The process of searching a subset of the whole dataset based on some kind of evaluation (or fitness) measure is called feature selection. Feature selection is a problem that has to be addressed in many fields. It is, particularly, a very important issue in remote sensing and GIS studies due to the availability of large amounts of spatial and satellite image data. As an evaluation function, several measures including Divergence, Transformed divergence, Bhattacharyya distance, Jeffries-Matusita distance, Hotelling's T² statistics and the Wilks' Lambda criterion can be used. In this study, the divergence measure is chosen to estimate the level of separability for each pair of features present in the dataset. Divergence was one of the first separability indices used in remote sensing and is still in use for processing of remotely sensed data (Goodenough *et al.*, 1978; Thomas, 1987; Mather, 1999). It is computed using the mean and variance-covariance matrices of the data representing feature classes. For two feature classes (*i* and *j*), the divergence between the classes is calculated according to the formula:

$$D_{ij} = \frac{1}{2} tr\left[\left(V_i - V_j\right)\left(V_j^{-1} - V_i^{-1}\right)\right] + \frac{1}{2} tr\left[\left(V_i^{-1} + V_j^{-1}\right)\left(M_i - M_j\right)\left(M_i - M_j\right)^T\right]$$

(1)

where $tr[\cdot]$ is the trace of a matrix, which is the sum of the diagonal elements of the matrix, V_i and V_j are the variance-covariance matrices for class *i* and *j*, and M_i and M_j are the corresponding mean vectors. In cases in which more than two classes are involved, average divergence is computed.

2.   Methodology

The results of the ANN classification process can be portrayed as a thematic map showing land cover types. The accuracy of the thematic map can be quantitatively assessed using fuzzy membership for each class. The major problem of such thematic maps when per-pixel classification is applied is that small isolated polygons may appear around field boundaries in the output image due to the presence of atypical and mixed pixels in the original image. As a consequence, the result image has a 'salt-and-pepper' look. In order to eliminate this effect, a low-pass filter is generally applied. A modal filter is used in this study for this purpose.

Since the accuracy of land cover classification is generally around 80% (Kavzoglu and Mather, 1998), reliable vector maps showing field boundary information could be extracted from such accurate classifications. In this study the result of the ANN classification, with an accuracy of around 90%, was vectorised to produce polygons, and the assessment of this map was carried out by comparing it with the digitised field boundary map derived from the Ordnance Survey 1:25,000 map. Figure 3 shows the methodology adopted in this study.

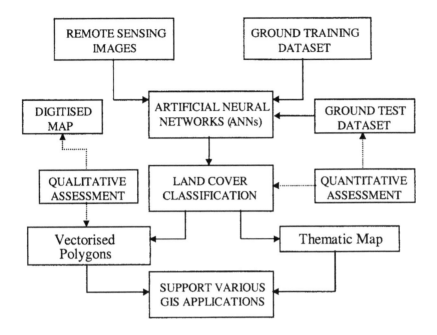

Figure 3. *Methodology for classifying land cover types and the derivation of polygonal vector map.*

3.   Results And Discussion

First of all, as it is not appropriate to include all the 24 bands available in the ANN processing, which could require considerable amount of time and make the network over specific, a sequential forward selection technique using the divergence separability index as the fitness measure was applied. The sequential forward selection technique (SFS) starts the search by determining the best individual band, then evaluates the remaining bands one at a time to find the second best band (i.e. the one that gives higher separability than other candidate bands). This process continues iteratively till a desired number of bands are selected. After a number of experiments on the optimum size of input layer, it was found that eight number of nodes, corresponding to eight spectral bands, would be sufficient for neural networks to learn the characteristics of the training data with around 90% overall classification accuracy

An ANN structure of 8-10-7 (8 indicates the number of inputs, 10 is the number of nodes in the hidden layer, and the number of output classes is 7) was found to be the optimum structure, following a series of experiments. These experiments were carried out based on the methodology proposed by Hirose et al. (1991), which was noted earlier. An overall accuracy of 90% and a kappa coefficient of 0.885 were achieved for the whole site.

The trained network was used to classify the test image and the result was filtered using a mode filter to reduce noise. The area of interest was extracted and used for further analyses. The extracted thematic map is shown in Figure 4.

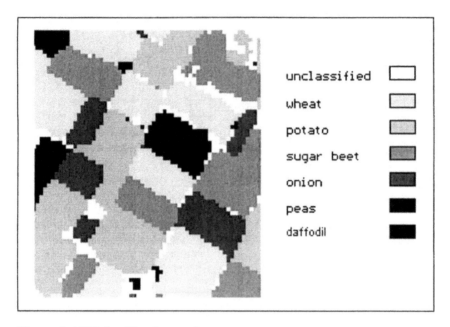

Figure 4. *ANN classification results.*

The boundaries of the land cover objects shown in the classified image (Figure 4) were then vectorised to produce a field boundary map (solid lines in Figure 5a). A buffer zone of 30 metre, corresponding to a pixel size, is produced for manually digitised polygons and overlaid on the vectorised boundary map. As can be seen, the vectorised land cover polygon boundaries are generally located inside the buffer zone. It is noted that when neighbouring fields include the same crop they are grouped together, producing a common polygon. It is also observed that the boundaries of the vectorised land cover polygons naturally show a 'zig-zag' pattern, as a result of the raster to vector conversion operation. This effect could be reduced by applying a line-generalisation algorithm, like that proposed by Jaafar and Priestnall (1999), or a line-simplification algorithm such as Douglas and Peuker (1978). In this study, a form of band method, considering maximum distance and azimuth difference between points as criteria, is applied to simplify the boundaries of fields using a C++ program developed specifically for the purpose. The result of this process is shown in Figure 5b.

Retrieval of the corresponding vectorised polygons from the classified image can be used as a guideline for updating the map produced much earlier than the satellite acquisition date. Such accurate polygonal maps might only be obtained using classifiers like ANNs because of the high accurate results produced. It could be argued that an uncertainty zone or a gap exists in the raster maps, separating the land cover polygons as a result of the inclusion of features such as headlands, hedgerows, or other boundary conditions. This effect can also be reduced by applying a low-pass filter to the classified image before vectorisation. Alternative way of avoiding this problem is to use GIS techniques to disregard such areas through buffer zoning the boundaries to be digitised. The size of the buffer zone should be defined by considering the pixel size of the image. It is a fact that at least one pixel is mixed between two neighbouring field pixels having different types of crops. Another finding in this study is that pixels left unclassified

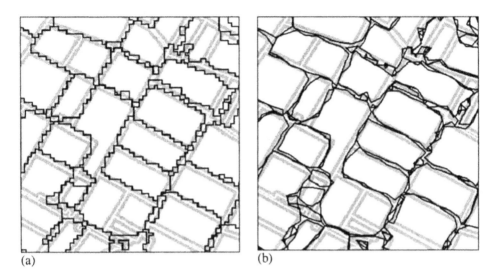

Figure 5. (a) *Vectorised polygon boundaries, extracted from classified raster image, overlaid onto 30 metre buffer zone of digitised boundaries,* (b) *The effect of zig-zagging due to the raster to vector conversion is eliminated using line simplification procedure.*

mainly corresponds to farm houses and other man-made objects, which can be observed by comparing the locations of farm houses in Figure 1 with classification results presented in Figure 4.

4.   Conclusion

We have shown that an accurate thematic map of an agricultural area can be produced using a supervised ANN classification. The boundaries of land cover regions (agricultural fields or groups of agricultural fields under the same crop) can be derived by raster to vector conversion procedures, and such information is useful in updating digital maps of agricultural areas. The utility of these procedures depends upon the geometrical accuracy of the image, the accuracy of the classification, and the quality of the raster to vector conversion. In this study, classification accuracies of 90% are achieved, partly as a result of careful design of the ANN structure.

As noted by Fischer (1999), neural networks provide not only novel and extremely valuable classes of data-driven mathematical tools for a series of spatial analysis tasks, but also an appropriate framework for re-engineering our well-established spatial data analysis techniques to meet the new large-scale data processing needs in GIS. Neural network approaches can be, therefore, seen as a potential tool for accurate analysis of large volume of spatial data present in a GIS environment. Using such techniques helps to provide better understanding and interpretation of spatial data.

Acknowledgements

The first author would like to thank the Turkish Government for funding his research. Research and computing facilities were made available by the School of Geography, The University of Nottingham. The authors are also grateful to Logica, UK for providing some of the SPOT images and to the Ordnance Survey of Great Britain for permission to reproduce the Ordnance Survey map.

References

Ardö, J., Pilesjö, P. and Skidmore, A., 1997, Neural networks, multitemporal Landsat Thematic Mapper data and topographic data to classify forest damages in the Czech Republic. *Canadian Journal of Remote Sensing*, Vol. 23(3), pp. 217-229.

Civco, D. L., 1993, Artificial neural networks for land-cover classification and mapping. *International Journal of Geographical Information Systems*, Vol. 7(2), pp. 173-186.

Douglas, D.H. and Peuker, T.K., 1973, Algorithms for the reduction of the number of points required to represent a digitised line or its caricature. *The Canadian Cartographer*, Vol. 10(2), pp. 110-122.

Fahlman, S. E. and Lebiere, C., 1990, The Cascade-Correlation learning architecture. *Advances in Neural Information Processing Systems 2*, edited by D. S. Touretzky, (San Mateo, California: Morgan Kaufmann), pp. 524-532.

Fay, T. H. and Miller, H. V., 1991, An automatic general-purpose linear feature extractor for digital multispectral imagery. *Proceedings of Marine Technology Society Conference (An Ocean Cooperative: Industry, Government & Academia)*, Vol. 1, pp. 489-494.

Fischer, M. M., 1999, Spatial analysis: retrospect and prospect. In *Geographical Information Systems: Principles and Technical Issues*, edited by P. A. Longley, M. F. Goodchild, D. J. Maguire, and D. W. Rhind. (New York: Wiley), pp. 283-292.

Goodenough, D. G., Narenda, P. M., and O'Neill, K., 1978, Feature subset selection in remote sensing. *Canadian Journal of Remote Sensing*, Vol. 4(2), pp. 143-148.

Gruen, A. and Li, H., 1995, Road extraction from aerial and satellite images by dynamic programming. *ISPRS Journal of Photogrammetry and Remote Sensing*, Vol. 50(4), pp. 11-20.

Hassibi, B. and Stork, D. G., 1993, Second order derivatives for network pruning: Optimal Brain Surgeon. In *Advances in Neural Information Processing Systems 5*, edited by S. J. Hanson, J. D. Cowan, and C. L. Giles. (San Mateo, CA: Morgan Kaufmann), pp. 164-171.

Hirose, Y., Yamashita, K. and Hijiya, S., 1991, Back-propagation algorithm which varies the number of hidden units. *Neural Networks*, Vol. 4, pp. 61-66.

Jaafar, J. and Priestnall, G., 1999, The importance of extracting buildings from LIDAR-derived DSMs. Proceedings of GIS Research UK 1999 (GISRUK'99), University of Southampton, pp. 73-79.

Jaafar, J., Priestnall, G. and Mather, P.M., 1999, Assessing the effects of grid resolution in laserscanning datasets towards the creation of DSMs, DEMs and 3D models. *Presented at the Fourth International Airborne Remote Sensing and Exhibition/21st Canadian Symposium on Remote Sensing*, Ottawa, Ontario, Canada.

Kavzoglu, T. and Mather, P.M., 1998, Assessing artificial neural network pruning algorithms, *Proceedings of the 24th Annual Conference and Exhibition of the Remote Sensing Society*, Greenwich, pp. 603-609.

Kavzoglu, T. and Mather, P.M., 1999, Pruning artificial neural networks: an example using land cover classification of multi-sensor images. *International Journal of Remote Sensing*, Vol. 20(14), pp. 2787-2803.

Kavzoglu, T., 1999, Determining optimum structure for artificial neural networks, *Proceedings of the 25ʰ Annual Technical Conference and Exhibition of the Remote Sensing Society*, Cardiff, pp. 675-682.

Le Cun, Y., Denker, J. S. and Solla, S. A. (1990), Optimal brain damage. In *Advances in Neural Information Processing Systems 2*, edited by D. S. Touretsky. (San Mateo, CA: Morgan Kaufmann), pp. 598-605.

Mather, P. M., 1999, *Computer processing of remotely-sensed images*. (Chichester: Wiley).

Merlet, N. and Zerubia, J., 1996, New prospects in line detection by dynamic programming. *IEEE Transactions on Pattern Analysis and Machine Intelligence*, Vol 18, pp. 426-431.

Paola, J. D. and Schowengerdt, R. A., 1995, A review and analysis of backpropagation neural networks for classification of remotely-sensed multi-spectral imagery. *International Journal of Remote Sensing*, Vol. 16(16), pp. 3033-3058.

Skidmore, A. K., Turner, B. J., Brinkhof, W. and Knowles, E., 1997, Performance of a neural network: mapping forests using GIS and remotely sensed data. *Photogrammetric Engineering and Remote Sensing*, Vol. 63(5), pp. 501-514.

Thomas, I. L., Ching, N. P., Benning, V. M., and D'aguanno, J. A., 1987, A review of multi-channel indices of class separability. *International Journal of Remote Sensing*, Vol. 8(3), pp. 331-350.

Wang, F., 1994, The use of artificial neural networks in a geographical information system for agricultural land-suitability assessment. *Environment and Planning A*, Vol. 26, pp. 265-284.

Werschlein, T. and Weibel, R., 1994, Use of neural networks in line generalisation. *Proceedings of the Fifth European Conference and Exhibition on Geographical Information Systems, Paris, France*, pp. 76-85.

5

On The Assessment Of The Spatial Reliability Of Thematic Images

C.A.O. Vieira and P.M. Mather

The estimation of the accuracy of a thematic classification derived from remotely sensed data is generally based on the confusion or error matrix. This matrix is derived by evaluating the performance of the classifier on a set of test data. The quantities derived from analysis of the confusion matrix include percent accuracy (total and per class), producer's accuracy, consumer's accuracy, and varieties of the kappa coefficient. None of these measures considers the spatial pattern of erroneously classified pixels, either implicitly or explicitly. Furthermore, each pixel in the image is assigned a unique ("hard") label, so that it is not possible to ascertain the reliability of the classification. In this paper we propose a methodology that specifically takes into account the spatial pattern of errors of omission and commission, and which presents the user with an indication of the reliability of pixel label assignments. Our methodology assumes that an accurate digital map of the spatial objects (fields, lakes, or forests) being classified, plus cultural features such as roads and urban areas are available. Such a map can be obtained by digitising a large-scale paper map of the study area, or via image processing.

Assuming also that the number of classes is appropriate to the problem at hand and to the scale of the image, it is likely that a substantial number of erroneous label allocations relate to mixed pixels located near or on field boundaries. A buffering operation is applied to the digitised boundary information in order to generate a mask. The width of the buffer is determined from a study of the location of the erroneously labelled pixels. To this mask are added regions not included in the classification, for example, roads and urban areas. In addition, spatial autocorrelation analysis was used to determine whether these remaining errors are spatially random or clustered in their distribution. Such information can help to refine the classification. Finally, a method using colour coding that allows the visualisation of the reliability of the classifier output is presented. These methods can be applied to most remotely sensed image types.

1   Introduction

The integration of remote sensing and Geographic Information Systems (GIS) in environmental applications has become increasingly common in recent years. Remotely sensed images are a cost-effective source for environmental GIS applications and, conversely, GIS capabilities are being used to improve image analysis and information extraction procedures and to allow their analysis in conjunction with other data. At the simplest level both remote sensing and GIS analysis can be conceived as consisting of three stages: measurement and sampling, the fitting of models or application of techniques to achieve some objective; and finally, *validation* of the results achieved. In this study emphasis is given to the third stage: *validation*.

Validation, in the form of accuracy assessment, is an essential part of most mapping, remote sensing and GIS database development activities. Performing spatial data analysis operations on data of unknown accuracy will result in a product with low reliability and restricted use in the decision-making process, while errors deriving from one source can propagate through the database via derived products (Lunetta *et al.*, 1991). The quality of data is a function both of the inherent properties of those data and the use to which they are to be put. Hence, knowledge of error levels is necessary if data quality is to be estimated. In spatial databases, it is equally true to say that knowledge of the spatial pattern of errors is necessary. Such knowledge can help identify areas of low reliability, which can help the analyst to devise strategies that specifically include such knowledge. Given the complexity of remotely sensed classified images, in which different regions of the thematic maps have varying accuracy, there is a need to assess the spatial reliability of the results, and these accuracy aspects should be clearly stated on a thematic map (Congalton, 1991).

The accuracy of a thematic classification derived from remotely sensed data is generally based on non-spatial statistics that summarise the characteristics of a confusion or error matrix, which relates the output of a classifier and known test data (Congalton, 1991, Lunetta *et al.*, 1991, Story and Congalton, 1986). These statistics include overall accuracy, individual class accuracy, user's and producer's accuracy and varieties of the kappa statistic. Although these measures are in widespread use, none of them considers the spatial distribution of erroneously classified pixels, either implicitly or explicitly. The aim of this paper is to describe methods of assessment of the performance of thematic images that explicitly include the spatial pattern of classification errors, and which presents the user with a visual indication of the reliability of the pixel label assignments.

2   Data

A SPOT High Resolution Visible (HRV) multispectral (XS) image (14 June 1994) of a region of flat agricultural land located near the village of Littleport (E. England) is used in this study as example of the variety of remotely sensed data, together with Field Data Printouts for summer 1994. These printouts are derived from official survey data supplied by individual farms, and provide details of the crop or crops growing in each field in the study area. The Field Data Printouts are used in the compilation of official agricultural statistics. On the basis of examination of the areas covered by each crop, the geographical scale of the study, and the spectral separability of the crops, seven crop categories were selected: potatoes, sugar beet, wheat, fallow, onions, peas and daffodil bulbs.

Image processing operations were performed using ERDAS Imagine (version 8.0) and the IDRISI GIS. Neural network application used the SNNS software. Some in-house programs were written to carry out specific procedures. Registration of the image to the

Ordnance Survey (GB) 1:25,000 map was performed using 17 ground control points and nearest neighbour interpolation. The RMS error was 0.462 pixels.

3 Methods

The type and sequence of procedures used in the analysis are outlined in Figure 1. Details are present in the following sub-sections.

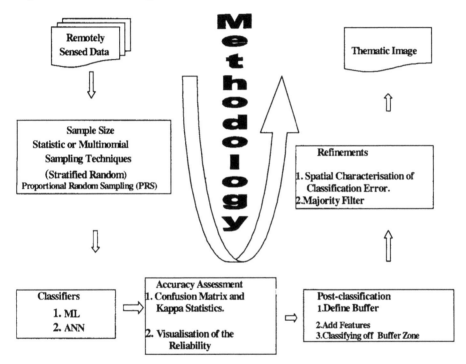

Figure 1. *An outline of the methodology followed in this study*

3.1 Sample Size and Sampling Techniques

Any analysis performed for accuracy assessment is statistically valid if an adequate and representative number of samples per map class are collected (Congalton and Green, 1999). However, the expensive costs of data acquisition require that the sample size be kept at a minimum to be affordable. Several researchers have published guidelines for choosing the appropriate sample size (e.g., van Genderen and Lock, 1977; Hay, 1979; Rosenfield *et al.*, 1982). Swain (1978) and Mather (1999) recommend that to generate a representative training sample for the multivariate case, there should be at least 30 times the number of discriminating variables (e.g. wavebands) per class, and preferably more. The majority of researchers have used equations which are based on the normal approximation to the binomial distribution to compute the sample size (Congalton, 1991). Those equations are correctly applicable to computing the sample size needed to estimate the overall accuracy of a classification where an analyst is interested only in the proportion of correctly classified samples and on some allowable error. However, when it is not simply a question of correct or incorrect pixel label (the binomial case), it is necessary to collect a sufficient number of samples to be able to represent adequately the confusion among all the classes in the study (Congalton and Green, 1999). One possible alternative is to use the multinomial distribution, as recommend by Congalton and Green (1999).

Let us assume an image classification problem with k mutually exclusive and exhaustive classes. In addition, let us also assume that Π_i, $i = 1,...,k$, be the proportion of the image area in the ith class. The sample size n (in a simple random sample) required to generate a valid error matrix can be obtained as follows: $n = B\Pi_i(1 - \Pi_i)/b_i^2$ where b_i is the absolute precision of the ith sample and B is the upper $(\alpha/k) \times 100^{\text{th}}$ percentile of the χ^2 distribution with 1 degree of freedom. Remember, α is the desired confidence level. Specifically, for our application: there are seven categories in our classification scheme ($k = 7$), the desired confidence level is 95%, the desired precision is 5% ($\alpha = 0.05$), and the class wheat covers 33% of the map area ($\Pi_i= 33\%$). The value B must be determined (or interpolated) from a Chi square table with 1 degree of freedom and significance level α/k (0.007142857). In this case the appropriated value for B is $\chi^2_{(1,0.992857)}= 7.348571$. Therefore, the calculation of the sample size is $n = 650$.

A total of 650 samples, or approximately 93 samples per class, should be taken to adequately fill a confusion matrix. It is very interesting that this value is very similar to the statistical recommendation 90/class (i.e., 30 x 3 bands).

If there is no prior knowledge of the values of the Π_i's, a conservative sample size calculation can be made assuming $\Pi_i = \frac{1}{2}$ and $b_i = b$ for $i = 1,..,k$. Thus, the minimum size n from the sample in this worst-case scenario can be computed using the formula: $n = B/4b^2$. If the simplified, worst-case scenario equation is used, then the class proportion is assumed to be 50% and the calculation is: $n = 735$. In this worst case, approximately 105 samples per class or 735 total samples would be required. Congalton and Green (1999) give a comprehensive review and development of these equations.

Another aspect of importance to any statistical analysis is the sampling strategy applied. A sampling strategy was selected to test classification accuracy in this study, which was carried out using stratified random sampling based on the reference image (ground truth). The reference image was generated at the same scale and projection system of the remotely sensed data. This sampling strategy was obtained using proportional random sampling (PRS), where the number of pixels per class is proportional to the areal coverage of the class, with a minimum of 105 pixels being selected. Ten sample sets, for each strategy (total 20 sample sets), were selected from the registered image set. Five independent sample sets were used to train the classifiers and five sample sets (selected at random) were reserved for validation. As might be expected in the use of the PRS strategy, the number of pixels per class is almost proportional to their occurrence in the study area. In other words, the most abundant classes have most pixels. Therefore, it can improve classification accuracy by weighting classes appropriately (Foody *et al.* 1995).

For a thorough discussion of sampling techniques, readers are referred to Congalton and Green (1999) and Jassen and van der Wel (1994).

3.2 Classification Techniques

A supervised classification procedure, using an Artificial Neural Network (ANN) and the Maximum Likelihood (ML) algorithm, were applied to the training sample sets. The

Gaussian Maximum Likelihood method is a well-known classification algorithm (Mather, 1999). A multi-layer perceptron using the back-propagation algorithm (Lippman, 1987; Bischof *et al.*, 1992; Benediktsson *et al.*, 1990) was used, with one neurone per spectral band in the input layer and an output layer with seven nodes, corresponding to the seven selected crop classes. The number of hidden layers (1) and the number of hidden nodes (10) were determined using the Hirose *et al.* (1991) procedure. The learning rate and momentum were kept constant at 0.2 and 0.9, respectively. In this research, the class allocations were performed using the modified form of the 'winner takes all' rule so that, for a given input, the label corresponding to the highest activation value in the output layer was selected, provided that this value equals or exceeds an threshold value of 0.7. This threshold was determined heuristically based on the authors' experience from previous work. If the highest activation of any output layer neurone is less than 0.7 then the corresponding pixel is labelled as "unclassified".

3.3 Accuracy Assessments and Visualisation of the Reliability

After the classification phase, standard accuracy measures derived from a confusion matrix were computed (Table 1) using a test data set based on the Field Data Printouts. In addition to these traditional methods, spatial assessment methods were introduced.

Table 1   Confusion Matrix

Classes	Reference Data							TOTAL	Users(%)	Z	K(cond)	Variance
	1	2	3	4	5	6	7					
1	112	15	0	1	18	0	0	146	76.7	20.352	0.751	0.001361
2	30	197	0	4	13	3	0	247	79.8	28.196	0.777	0.000759
3	11	1	1012	0	0	2	2	1028	98.4	144.143	0.973	0.000046
4	1	0	0	90	2	0	2	95	94.7	39.626	0.945	0.000569
5	0	4	0	2	17	0	0	23	73.9	7.859	0.733	0.008709
6	0	0	0	0	0	113	8	121	93.4	39.129	0.93	0.000565
7	0	0	0	0	0	6	64	70	91.4	26.422	0.911	0.00119
Unrecognised	199	247	15	16	75	25	29	606				
TOTAL	353	464	1027	113	125	149	105	2336	OVERALL	Z	Kappa	Variance
Producers(%)	31.7	42.5	98.5	79.6	13.6	75.8	61		88.70%	60.91	0.604	0.000098

The measures based on the confusion matrix were: overall accuracy, individual class accuracy, producer's accuracy and users' accuracy. The calculations associated with these measures are described in standard textbooks (e.g., Mather, 1999). The kappa coefficient, conditional kappa for each class and test Z statistics - all of them widely used statistic derived from the contingency matrix - were also computed (Congalton and Green, 1999).

In addition, a pairwise test for testing the significance of the classifiers (represented here by their respective confusion matrices), was performed utilising the Kappa coefficients. These results are summarised in the form of a *significance matrix*, in which the major diagonal elements indicate if the respective classification result is meaningful. In this single confusion matrix case, the Z value can be computed using the formula:

$$Z = Ka \big/ \sqrt{\mathrm{var}(Ka)} \qquad\qquad (1)$$

where Z is a standardised and normally distributed test statistic and *var* is the large sample variance of the kappa coefficient K. If $Z \geq Z_{\alpha/2}$, the classification is significantly better than a random allocation, where $\alpha/2$ is the confidence level of the two-tailed Z test and the degrees of freedom are assumed to be infinity.

On the other hand, the off-diagonal elements give an indication, again if $Z \geq Z_{\alpha/2}$, that the two independent classifiers are significantly different. The formula used to test for significance between the two independent kappa coefficients is:

$$Z = |Ka_1 - Ka_2| / \sqrt{\mathrm{var}(Ka_1) + \mathrm{var}(Ka_2)} \qquad (2)$$

where the Ka_1 and Ka_2 are the two Kappa coefficients being compared (Congalton and Green, 1999).

The methods discussed above for quantifying error in remotely sensed data are very important and widely understood. However, they do not consider the spatial distribution of misclassified pixels.

One possible way to characterise the spatial distribution of the errors in a thematic classification is by generating a *distance image* (see Figure 2(a)) showing the deviations of individual pixels from the means of the classes to which they have been assigned (Jupp and Mayo, 1982). Either the Euclidean distance or the Mahalanobis distance can be used. The former, however, implies spherical clusters in feature space, while the latter takes into account the covariance between the features on which the classification is based. The individual distances are scaled onto a 0-255 range, and displayed as a grey scale image. Darker pixels are spectrally "nearer" to their class centroid (in the sense of statistical distance), and are thus more likely to be classified correctly. On the other hand, pixels with higher distance values are spectrally further from the centroid of the class to which they were assigned, and are thus more likely to be misclassified. A threshold can be

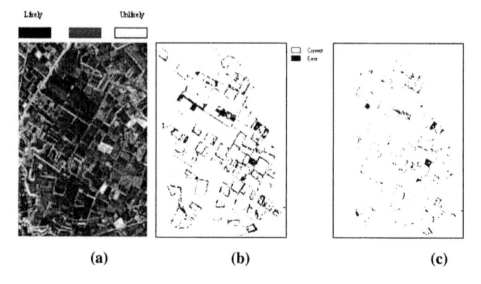

(a)   (b)   (c)

Figure 2 *Spatial characterisation of classification errors using thematic image generated by ANN (385 x 285 pixels). (a) The distance image (b) Error binary image before buffering operation and (c) Error binary image after buffering operation.*

applied to the distance image to identify those pixels that are most likely to be misclassified. By examining the spatial distribution of such pixels in Figure 2(a) we can make a number of observations. It is apparent that misclassified pixels are spatially correlated. These correlation effects are considered below in more detail, and are probably due to the presence of mixed pixels at field boundaries (Flack, 1995). The variation in the reflectance spectrum is caused, most probably, by variations in soil type within a field, or to the effects of crop management practices such as the use of fertilisers.

An alternative way of looking at the spatial distribution of the errors present in a classified image is by directly comparing thematic images with their respective ground truth maps. One of the products of this comparison should be a binary error image (Figure 2(b)) in which each point takes the value 0 (correctly labelled) or 1 (erroneously labelled). Another product of this comparison is the *difference image*, in which not only one but also two different thematic images are compared with their respective ground truth maps. If for the two thematic images corresponding pixels are correctly labelled the difference image receives a code, for instance 11. On the other hand, if both corresponding pixels are wrongly labelled the difference image receives another code 00. Two other codes can be assigned to a pixel in the difference image. If for one thematic image the pixel is correctly labelled and for other wrongly labelled the difference image receives another code (e.g., 10) and vice-versa the image receive the code 01. As usual, a separate colour is assigned to each code. This kind of representation allows the visual appreciation the spatial distribution of erroneously classified pixels in both thematic images at once. Figure 3 shows an example of difference image.

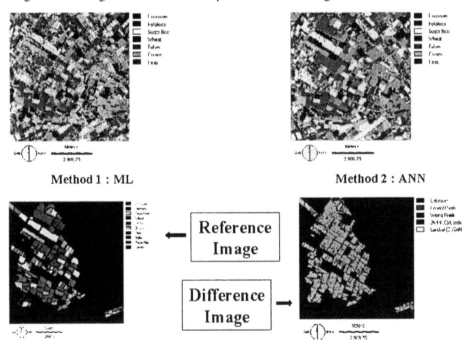

Figure 3. *An example of a difference image.*

The Mahalanobis Distance (MD) can be used to compute a measure of reliability of a pixel's label. The MD is transformed into probabilities using the following formula (Xu *et al.*, 1992):

$$P = MD^{-1} / \sum_k MD^{-1} \tag{3}$$

where MD is the Mahalanobis Distance between the pixel and the mean pixel values (or prototypes) of each class. A "new value" (NV) is derived by multiplying the assigned class label by 10 and then adding the probability computed from the Mahalanobis Distance by 10, so that NV= (class label * 10) + (P * 10). For example, a pixel allocated to class 3 with probability 0.64 gives an NV of 36. Pixels in class 1 therefore have a range of NV of 10 to 19 inclusive; class 2 takes the range 20-29 inclusive, and so on. As usual, a separate colour is assigned to each class. These within class levels are also assigned separate shades of that colour, so that each class is represented by five shades of the given colour (see Figure 4(a)). This kind of representation allows the visual appreciation of the degree of accuracy of the classified crop. A contour representation of the reliabilities can also be used (Figure 4(b)). These types of representation help the user to identify portions of the thematic map that have reduced reliability. Although the final map may look uniform in its accuracy, it is actually a representation assemblage from diverse image processing procedures and refinements. It is important for the user to known how these accuracies are spatially distributed in the image through a thematic reliability map.

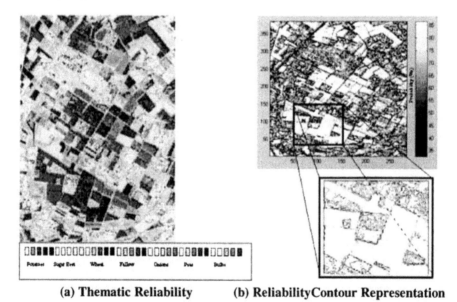

(a) Thematic Reliability   (b) ReliabilityContour Representation

Figure 4   *Representation of the Reliabilities Using ANN (385 x 285 pixels).*

3.4   Post-classification Procedures: Defining Buffer Zones

Our methodology requires that an accurate digital map showing the spatial objects (fields, lakes, forests, etc.) being classified, plus cultural features such as roads and urban areas, is available at an appropriate scale. We digitised a 1:25,000 paper map of the study area (Figure 5). It may be possible to obtain some boundary or edge information via image processing (Flack, 1995). Assuming also that the number classes are appropriate to the problem at hand and to the scale of the image, it is observed (Figure 2) that a substantial number of erroneous label allocations refer to mixed pixels located near, or on field boundaries.

A buffering operation was therefore applied to the digitised boundary information in order to generate a mask. The width of the buffer was determined from a study of the location of the erroneously labelled pixels. In this specific application, two different buffer widths were selected. The first had a width of 50 metres from all the cultural features (roads, railways, and so on) and the second had a width of 25 metres (field boundaries). To this mask were added regions not included in the classification, for example roads, railways and drains.

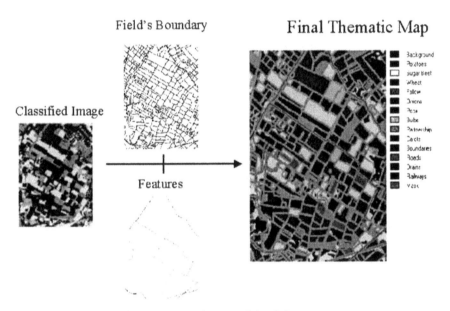

Figure 5 *Post-classification procedures and final thematic map.*

3.5   The Spatial Characterisation of Classification Error: Refining the Classification

Finally, we looked at the spatial distribution of the remaining errors to determine whether they are random or clustered in their spatial distribution.

Spatial autocorrelation measures the tendency of samples to cluster spatially (Cliff and Ord, 1973). According to Ebdon (1997), strong spatial autocorrelation means that adjacent values or ones that are near to each other, are strongly related, while values that are simply arranged at random over the surface should show no apparent spatial autocorrelation. Congalton (*1988*) investigated the spatial autocorrelation of remotely sensed data and determined its affects on accuracy assessment.

Ebdon (1997) presents a simple method of quantifying the degree of clustering or dispersion, in the spatial arrangement of areas characterised by the values 0 or 1. The method uses Join Count Statistics (JCS). In this approach, the number of joins between the areas, for instance black/white areas, is computed. If like areas are dispersed or random (null hypothesis), there will be relatively many black/white joins. If like areas are clustered together, there will be relatively few joins. In other words, a high spatial autocorrelation implies that the null hypothesis – that spatial error distributions are randomly distributed – is rejected. By measuring the spatial autocorrelation of the classified image against the reference image (ground truth) we may infer the existence of spatial pattern within the distribution of classification errors.

Given a binary image, such as the one shown in Figure 2(b), and using a non-free sampling approach (sampling without replacement as suggested by Congalton (1988)), Vieira and Mather (1999) performed Join Count Statistics analysis to identify and quantify the patterns of error within the binary image before and after the buffering procedures. They found that the z calculated value is less than the critical value obtained from tables of the standard normal deviate, so the null hypothesis can be rejected. The observed arrangement of black and white pixels (Figure 2(b)) is thus very unlikely to have occurred by chance; error pixels can be said to be 'significantly clustered' at the 0.05 significance level. Observing this result, it may be the case that registration error between the ground truth and the classified image exaggerated the effect of boundary mixing. In addition, Vieira and Mather (1999) used the same procedures to compute the z value in the error image after applying the buffering procedures (Figure 2(c)). As the calculated value z is of the correct sign (negative) and less in absolute value than the critical value, the null hypothesis could not be rejected and, therefore, the pattern did not depart significantly from the randomness at the 0.05 level. Once the boundary pixels were removed, the binary error image was left with uncorrelated (random) misclassifications. They argued that it could be due to spectral response variations within the fields, caused either by a variety of mechanisms (such as: differences in soil fertility, localised crop stress, and differences in the crop growth) or an insufficient training sample has been taken to represent each crop type.

Visual analysis of the spatial distribution of the remaining errors can help to refine the classification process. One possible refinement to the thematic image could be applied by employing a majority filter (contextual information), under the assumption that pixels of a given class are likely to be surrounded by pixels of the same class (Mather, 1999). Another refinement should be by assuming that the minimum spatial unit of interest is the field. Therefore, we could transform the thematic classified image to a field-based classification, considering also the majority of label in each field.

4   Results And Discussions

From the evaluation of the proposed post-classification procedures, it is also important to determine quantitatively whether the buffering operation contributes to improved classification accuracy of the classifiers. To determine this, a pairwise test, statistics for testing the significance of the classifiers (before and after the buffering operation), are summarised in form of a significance matrix, Table 2, for accuracy assessment were obtained using Maximum Likelihood (ML), Artificial Neural Networks (ANN). The proportional random sampling (PRS) was used for this experiment, since it presents better results to the data set used (Vieira and Mather, 2000).

Comparing the classifier performances (off diagonal elements), as expected, there are "positive" significant improvements (see grey classifier pairs) for the individual classifier's performances (e.g., for the ML, $Z = 4.50 > 1.96$ and for the ANN, $Z = 4.87 > 1.96$). Overall, the majority of experiments suggest that the use of a buffering strategy is able to improve significantly the classification performance. The performance of the classifiers were considerable improved when the mask procedure was applied, since the

Table 2 *Significance matrix for comparison among the classifiers using kappa analysis. The table also presents the Kappa coefficients and variance for each classifier. The Z values (along the major diagonal) and the Z values (off diagonal elements) were computed using Equations (1) and (2) respectively. Shaded classifier pairs indicate significant improvements in the performance of the classifiers at 95% confidence level (Z critical value = 1.96).*

Classifiers	ML	ANN	ML(Buffer)	ANN(Buffer)
KAPPA	0.583	0.67	0.651	0.742
VAR	0.0001	0.00011	0.000116	0.000106
ML	55.09			
ANN	5.80	63.03		
ML(Buffer)	4.50	1.26	60.44	
ANN(Buffer)	10.77	4.87	6.11	72.07

border training patterns appear to be less distinct or more ambiguous. This supports the view that the ambiguous cases may degrade classification training.

Finally, the goal of most remote sensing investigation is to produce a product (see final thematic map, Figure 5) that will quickly and accurately communicate important information to the scientist and decision-maker. The final product may take a number of forms and should also include statistical (table 1) and spatial error summaries (Figure 2), reliability maps (Figure 4), which will give to the final user a better understanding of the potential error sources associated with remote sensing data products.

5   Conclusions

This paper introduces spatial techniques for assessing the accuracy of classifications based on remotely sensed data. The techniques take into account the spatial pattern of errors of omission and commission, and present the user with an indication of the reliability of pixel label assignments. A considerable amount of research and development needs to be accomplished before the spatial characterisation of classification errors associated with remote sensing can be adequately reported in standardised format and legends. An analysis of the spatial incidence of classification error indicates a distinct spatially correlated pattern of error. These errors are usually significantly clustered, with a high proportion of erroneously labelled pixels occurring near boundaries. The use of a buffering procedure seems to overcome this problem. The study of the spatial distribution of the remaining errors, after the buffering operation, also helped in the classification refinement process. The use of this pilot classification approach, where the cultural features (such as roads, rivers and urban areas) are previously available, provides a far more appropriate and accurate interface between a remotely sensed data and a GIS.

Acknowledgements

Mr. Vieira's research is supported by CAPES (Brazilian Research Council). We are grateful to Logica PLC and SPOT Image for permission to use their images. We are also grateful to Mr. T. Kavzoglu for providing software to allow us to display the contour representations using the MATLAB system. Computing facilities were provided by the School of Geography, The University of Nottingham.

References

Benediktsson, J. A., Swain, P. H. and Ersoy, O. K., 1990, Neural network approaches versus statistical methods in classification of multisource remote sensing data. *IEEE Transactions on Geoscience and Remote Sensing,* Vol. 28, pp. 540-552.

Bischof, H., Schneider, W. and Pinz, A. J., 1992, Multispectral classification of Landsat images using neural networks. *IEEE Transactions on Geoscience and Remote Sensing,* Vol. 30, pp. 482-490.

Cliff , A. D. and Ord, J. K., 1973, *Spatial Autocorrelation.* Pion Press, London.

Congalton, R. G., 1988, Using spatial autocorrelation analysis to explore the errors in maps generated from remotely sensed data. *Photogrammetric Engineering and Remote Sensing,* 54(5), pp. 587-592.

Congalton, R. G. and Green, K., 1999, *Assessing the Accuracy of Remotely Sensed Data: Principles and Practices.* New York: Lewis Publishers.

Congalton, R. G., 1991, A review of assessing the accuracy of classifications of remotely sensed data. *Remote Sensing of Environment*, Vol. 49, No. 12, pp. 1671-1678.

Ebdon, D., 1997, *Statistics in Geography*, Second edition, Alden Press, Oxford.

Flack, J., 1995, *Interpretation of Remotely Sensed Data Using Guided Techniques*, Ph.D. Dissertation, School of Computer Science, Curtin University of Technology, Western Australia, November.

Foody, G., 1995, Training pattern replication and weighted class allocation in artificial neural network classification. *Neural Computing and Applications*, Vol. 3, pp. 178-190.

Gong, P., and Howarth, P. J., 1990, An assessment of some factors influencing multispectral land-cover classification. *Photogrammetric Engineering and Remote Sensing,* Vol. 36 (5), pp. 597-603.

Hay, A. M., 1979, Sampling designs to test land-use map accuracy. *Photogrammetric Engineering and Remote Sensing*, Vol. 45 (4), pp. 529-533.

Hirose, Y., Yamashita, K. and Hijiya, S., 1991, Back-propagation algorithm which varies the number of hidden units. *Neural Networks*, Vol. 4, pp. 61-66.

Jassen, L. L. F., and van der Wel, F. J. M., 1994, Accuracy assessment of satellite derived land-cover data: a review. *Photogrammetric Engineering and Remote Sensing*, Vol. 48, pp. 595-604.

Jupp, D. L. B., and Mayo, K. K., 1982, The use of residual images in Landsat image analysis. *Photogrammetric Engineering and Remote Sensing*, Vol. 60, No. 4, pp. 419-426.

Lippmann, R. P., 1987, An introduction to computing with neural nets. *IEEE ASSP Magazine*, April, 4-22.

Lunetta, R. S., Congalton, R.G., Fenstermaker, L. K., Jessen, J. H. and McGwire, K. C., 1991, Remote sensing and geographic information system data integration: error sources and research issues. *Photogrammetric Engineering and Remote Sensing*, Vol. 57, No. 6, pp. 677-687.

Mather, P. M., 1999, *Computer Processing of Remotely-Sensed Images: An Introduction.* Chichester: John Wiley and Sons, Second edition.

Rosenfield, G. H., Fitzpatrick-Lins, K., and Ling, H., 1982, Sampling for thematic map accuracy testing. *Photogrammetric Engineering and Remote Sensing*, Vol. 48 (1), pp.131-137.

Story, M. and Congalton, R. G., 1986, Accuracy assessment: A user's perspective. *Photogrammetric Engineering and Remote Sensing*, Vol. 61, pp. 391-401.

Swain, P. H., 1978, Fundamentals of pattern recognition. In *Remote Sensing: the Quantitative Approach*, edited by P. H. Swain and S. M. Davis, (New York: McGraw-Hill), 136-187.

Vieira, C. A. O. and Mather, P. M., 1999, Assessing the accuracy of thematic classifications using remotely sensed data. In *Proceedings of the 4th International Airborne Remote Sensing*, Ottawa, Canada, 21-214 June, pp. 823-830.

Vieira, C. A. O. and Mather, P. M., 2000, Visualisation of Measures of Classifier Reliability and Error in Remote Sensing. In *ICG Accuracy 2000*, Heuvelink, G.B.M. and Lemmens, M.J.P.M. (Editors), Amsterdam: Delft University Press, The Netherlands, pp. 701-708.

Xu, L., Krzyzak, A. and Suen, C. Y., 1992, Methods of combining multiple classifiers and their applications to handwriting recognition. *IEEE Transaction on Systems, Man, and Cybernetics*, Vol. 22(3), pp. 418-435.

Section 2

Manipulating the information: Tools, Visualisation and Navigation

6

A Triangle-based Carrier for Geographical Data

Morten Dæhlen, Morten Fimland and Øyvind Hjelle

In this paper we discuss a framework for representation and maintenance of multi-resolution terrain and network data. We use a triangle-based representation of the terrain surface, and representations of curve networks (river systems, roads, railroad, etc. are integrated into the triangle-based representation. We introduce a data model for the representation and we discuss issues connected to the construction of the integrated representation of data. Several examples are provided.

1 Introduction

The motivation for this paper is the large amounts of fine-resolution geodata available and the observation that terrain data may constrain the curve network and that the curve network may constrain the terrain. It is also a fact that these two data types are heterogeneous and inconsistent with respect to each other and between different resolution levels. Our aim is to create a framework to handle integration and consistent maintenance of curve network and terrain data, to enable efficient extraction of data at required resolution level and to ensure efficient and compact representation of the data.

Hierarchical models, also referred to as level-of-detail models or multi-scale models, are widely studied in the literature; see Herbert and Garland (1997) and references therein. Within geographical information systems (GIS) these models are particularly interesting. Multi-scale models are used to obtain compact storage of huge data sets, fast retrieval of data at different resolution and efficient storage of digital maps at different scale. There are several reasons for this. In a multi-scale model only differences are stored which in most cases requires less storage space. Moreover, data points can be removed based on some prescribed tolerance and for curve networks and terrain surfaces this is often the case. A multi-scale model is also constructed so that data can be retrieved at different resolutions, and in general this speeds up the data retrieval procedures. For example, over the Internet we can observe this through so-called progressive transmission and presentation of images and geometric objects. In some cases level-of-detail models also form a basis for cartographic generalisation. Throughout the last decade several papers are written on the subject of introducing multi-scale representations into GIS; see Ware and Jones (1992), Jones et al (1994), De Floriani and Puppo (1995) and Cignoni et al (1995) and references within these papers. Multi-scale models are also used in other areas, like computer aided geometric design, multi-grid methods for solving partial differential equation and in signal and image processing [Herbert and Garland (1977)]. In this paper we investigate the integration of curve networks into hierarchical triangle-based terrain surfaces. The triangle-based terrain representation is used as a carrier of curve networks like rivers, roads, etc. The underlying triangulation is also used in order to handle and maintain the topological relations between different curve networks or branches of the curve networks. In this paper we introduce a hierarchical model, which combines terrain surfaces with curve networks. Our goal is to produce a consistent hierarchical representation of terrain surfaces and curve network objects like rivers and roads lying on the terrain surfaces. We combine a hierarchical triangle-based representation of the terrain surface with a planar graph representation of the curve networks. Moreover, we discuss some issues connected to the implementation of the model and functionality for constructing and maintaining the hierarchical model.

The outline of the paper is as follows: In order to motivate for the work presented in later sections, we will in Section 2 discuss some major concepts in the proposed framework for multi-level representation of geographical data and we introduce the *geodata carrier*. In Section 3 we introduce a data model for multi-level representation of curve networks integrated into terrain models. A triangle-based representation of the terrain surface is used as an underlying framework for the integration. Related work can be found in Jones et al (1994). The multi-level structure is obtained by applying a specialised decimation method on the triangulation taking into account the structural relations between the curve networks on the terrain surface. The decimation method is based on the edge ordering technique presented in Daehlen and Finland (1999). Examples are also presented and discussed in Section 3. In Section 4 we give a brief comment on the operational issues of the framework. We summarise the paper in Section 5 together with some remarks on future work.

2   Framework

Geographical data, in the following referred to as geodata, plays an important role in many applications. With emphasis on a multi-level representation, we are in this section concerned with the construction, maintenance and distribution of scale-independent terrain models over huge areas. Moreover, we focus on issues connected to the integration of geographical curve network data into the terrain surface. In order to explain the various challenges we introduce the concept of a *geodata pipeline*. The geodata pipeline can be regarded as a basic processing system, either as an integrated module in a (geographical) information system or as a separate system for pre-processing geodata for further use.

Figure 1 shows a high level structure for the *geodata pipeline* designed for constructing models which integrate elevation and curve network data. Typical input data are scattered measurement data, grid data, contours, cartographic data and polygonal (line) data representing ridges, roads, rivers etc.

The *geodata carrier* is a triangle-based terrain surface defined over some suitable portion of the Earth's surface – the domain over which the terrain surface is defined. For huge volumes of data over large areas it is often convenient to decompose the domain into a set of non-overlapping tiles or subdomains. We will only investigate one domain in this paper. A strategy for domain decomposition can be found in [Sevaldrud (1999)]. The representation of the triangulation must be hierarchical in the sense

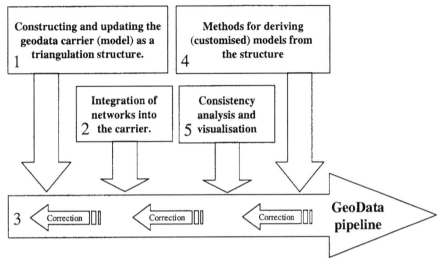

Figure 1. *The geodata pipeline based on a hierarchical representation of triangle-based terrain surfaces and associated curve networks.*

that a triangle-based terrain model can be extracted at different resolutions (or scales). Each triangle-based terrain model in the hierarchy can be a standard Delaunay triangulation [Nielson (1998)] or some other suitable triangulation of the parts of the underlying data set. *Constrained triangulation* [De Floriani and Puppo (1992)] is important since we will impose curve networks into the terrain. Constraint triangulation means that we force edges to be a part of the triangulation. Various types of pre-processing can be regarded as part of the construction and updating procedures, e.g. sorting points before constructing the hierarchies. The model construction implies building triangulations from input data into a hierarchical representation. The following items correspond to the numbers in the boxes in Figure 1.

1. The underlying data model, or the geodata carrier, holds the triangle-based terrain surface and its associated curve networks. The data model is represented as a graph that is not necessarily planar, and the graph can be realised as the terrain surface, curve networks on the surface, or relations between the curve networks on the surface. Hence, the two major entities of a graph, nodes and edges, include features, which classifies the entities into groups. The algorithms for constructing and updating the graph operate on the underlying data model.

2. The integration of curve networks into the geodata carrier (or data model) implies the construction of a planar graph based on some suitable data. Moreover, the planar graph (curve network) is integrated in the carrier. Conceptually, this can be done in two ways. As indicated in Figure 1, integration of curve networks is done after the construction of the carrier, which in most cases will impose corrections on the carrier. This is indicated in the figure and discussed in item 3 below. In general the integration of curve networks will impose constraints on the triangulation. Alternatively, the curve networks can be integrated directly when the carrier is being built. We will focus on the first approach since curve networks can be with or without elevation values.

3. The geodata carrier will in general be a set of terrain models at different resolutions. The curve networks are rivers, lake-boundaries, roads, or some suitable partition of the earth's surface, e.g. forest stands. When introducing a curve network we will in general have to change the terrain. A typical example arises when we introduce a river curve network, which of course is monotonically decreasing from the "mountain" to the "ocean". In this case we face the problem of correcting the terrain. Some correction can be detected and corrected automatically, however, in general this is a difficult task, which we have put up as a separate issue in item 5. As the data are refined through the pipeline, a system like this requires methods for handling corrections that are detected throughout the processing of the data.
4. From an application point of view it is important to provide functionality for extracting/deriving customised information from the underlying data model. Typically, a cartographer selects an area of interest for extracting information about road and river curve networks together with their spatial relations. Other derived models can typically be surface triangulations at a requested level of detail selected from a hierarchical representation, or we might only be interested in the topological representation of a river curve network.
5. In order to detect and correct inconsistencies in the data, a visualisation/graphics environment with interactive facilities is required.

The challenges behind item 4 and 5 are not discussed in this paper. Moreover, a database system is needed to store the data, and we assume that a suitable database is available. In the examples presented later we are using a fairly straightforward file structure for storing the graphs and the triangulation. In this paper we focus on the major functionality behind the modules marked 1, 2 and 3 in Figure 1 and how these modules are linked.

In order to implement the framework some basic technologies have to be available. We need software that efficiently can construct and handle large triangulations including various methods for constructing a triangulation, e.g. inserting edges into the triangulation. For this purpose we may use a triangulation library like SISCAT [SINTEF (1995)]. We can use the software such that a standard one-level triangulation is regarded as a special case of the multi-level triangulation. Moreover, the software has been extended so that the curve networks can be integrated using methods for constrained triangulation [De Floriani and Puppo (1992)]. A singular point can be regarded as a degenerated curve network or graph. Finally, storage requirements for the adjacency information between topological entities as vertices, edges and triangles should be minimal, but sufficient for carrying out topological operations as fast as possible. With respect to these issues, the most important is to enable *representation*, and not necessarily *construction*. The construction of the triangulation or the graph can be regarded as an application issue and several methods are available, see [Nielson (1998)].

3   Data Model

In this section we introduce the data model which integrates the triangle-based hierarchical terrain model with a set of curve networks. The model is finally represented as one single graph. When introducing hierarchical triangulations we assume that a hierarchical structure consists of a number of independent triangle-based surfaces at different level of resolution. This assumption implies that we can use standard data structures for representing triangle-based surfaces over some suitable domain. The issue of combing the topological information of different levels of the hierarchical triangulation is interesting due to the possibility of reducing storage requirements. However, this problem is due to further studies.

There are many possible topological structures, or data structures, for representing triangulations on computers. On the one side the data structure must be chosen in view of the needs and requirements of the actual application in mind. On the other hand we must design structures which are efficient with respect to the basic operations which we want to perform, e.g. traversing the triangulation. When analysing different data structures we always face a trade-off between storage requirement and efficiency of carrying out topological and geometric operations. For example, for visualisation purposes we need a data structure with fast access to data and sufficient topological adjacency information for traversing the topology of the triangulation when extracting sequences of triangles for the visualisation system. This will normally require more storage than a data structure used only for storage purposes. In a real world application we might need more than one data structure and thus tools for mapping one data structure to another are important [Nielson (1998)].

3.1 Graphs

Behind the model we can use graph theoretic concepts for representing the topology of the triangulations and the curve networks. Thus, we will make a clear distinction between the topological elements such as vertices, edges and faces, and the geometric information which we associate with these topological elements, Geometric information are points, curves (or straight line segments) and surfaces in 3D space. By considering triangulations and curve networks as graphs, we can benefit from an extensive mathematical theory and a variety of interesting algorithms operating on graphs. There is a rich literature on graphs, algorithmic graph theory and applications of graphs; see [Wilson (1985), Behzad (1979), McHugh (1990)] and references therein.

The graph representation of a triangulation, which represents a subdivision of an open surface into triangles, will be a *planar graph*, G(V,E), where V is a set of vertices and E is a set of edges connecting vertices in V. Loosely speaking, a planar graph is a graph that can be embedded, or drawn, in the plane such that no two edges intersect except at their incident vertices.

(a)   (b)

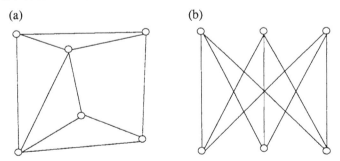

Figure 2. *A planar embedding of a graph in (a), and a non-planar graph in (b).*

The graph shown in Figure 2(a) is planar according to this definition. Figure 2 (b) shows a graph that is non-planar. Depending on the restrictions we decide to put on triangulations, we can make restrictions on G such that traversing operations become faster. The trivial restriction is that each edge and vertex of the graph is a member of a minimum *edge-vertex cycle*, or a *face cycle*, with three vertices and three edges. This implies that we will only model faces that are triangles. We might also decide to restrict G further and model *regular* triangulations only, i.e. the domain is without holes and connected edge by edge.

To follow the speed-up issue further, we may choose a graph representation that reflects data structures based on the notion of *half-edge* as the basic topological element. Each edge in E, connecting two vertices in V, will then be replaced by two *ordered* edges pointing in opposite directions. The graph $G(V,E)$ will then be a *planar ordered multigraph*, or more specifically, a *planar bidirected graph* where face cycles are pairwise disjoint. We can also use the term *planar embedded bidirected graph* for a graph representing the half-edge structure in Figure 4, since we assume that the graph objects, vertices and edges, are embedded with geometry.

Representing hierarchical structures in combination with curve networks requires graph representations that in general are *non-planar* (c.f. Figure 2 and the definition of planar graphs above). While triangulations or simple curve networks can be regarded as planar graphs when viewed separately, the union, when integrated into one single graph, is not necessarily planar. For example, a road curve network imposed into a triangulation will result in a planar graph, while a graph representing a curve network connecting cities combined with the terrain graph may not be planar. Again, we might regard this as an application issue. However, the generic tools must enable representation and manipulation of these structures.

It is an open question how (or if) these graph models should be implemented. The needs may differ from one application to the other and it might be difficult to find a sound unified design covering the needs of all applications. However, a generic graph library such as LEDA (1988) would provide many useful tools and data structures for triangulations and curve network software. In the first instance, the most important task would be to establish flexible generic data structures for graph representations without sophisticated graph algorithms operating on these structures.

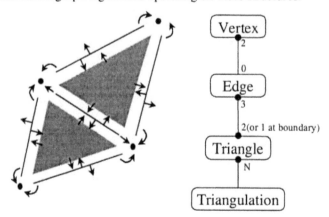

Figure 3. *Pointers in the triangle-edge topology for triangulations.*

3.2   Triangulation

The first data structure described here has been implemented in many triangulation libraries. The topology is represented as a fixed pointer structure as depicted in Figure 3. A triangle has references to its three edges in counter-clockwise order, and an edge has references to two nodes. In addition, there are references from edges to incident triangles. A triangulation consists of a list of all triangles.

We observe that the edges are not oriented when viewed separately. When viewed from a triangle they can be considered as oriented since a triangle refers to its three edges in counter-clockwise order. It can be shown that that the number of pointer fields in such a triangulation is approximately twenty times the number of vertices. We will use a topological structure that carries out topological operations more efficiently.

The notion of *half-edge* as the basic topological entity for boundary based topological representations was introduced by Weiler (1985). The principle is to split each edge into two directed half-edges each of which are oriented opposite to the other as shown in Figure 4. Hence, we can think of a half-edge as belonging to exactly one triangle, and the three half-edges of a triangle can be oriented counter-clockwise around the triangle. There are many possible data structures that can be derived from this concept, for example, *vertex-edge* and *face-edge* data structures, see Ware and Jones (1992).

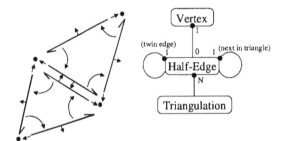

Figure 4. *Pointers in the half-edge topology for triangulations.*

Each half-edge has a reference to the node it starts from, a pointer to the next half-edge belonging to the same triangle (counter-clockwise), and a pointer to its twin edge belonging to another triangle. We have a list of "triangles", but each triangle is only represented as one of its half-edges. It can also be shown that we have exactly the same storage requirement as in the triangle-edge structure shown in Figure 3.

Basic topological operations for traversing the topology will be faster with this data structure because the half-edges are oriented. But we also loose one important property. There are no triangle objects in the internal structure. Thus, when visiting the three half-edges in a triangle we do not know which represents the triangle. This might be necessary, for example when removing a triangle from the triangulation. On the other hand, in an actual implementation we might need extra information in the topological elements anyway, for example bit-fields, so then its sufficient with one bit to indicate whether a half-edge represents a triangle or not.

3.3 Curve networks

In the following we will describe the structure for *curve networks*. Figure 5 shows a standard representation of a curve network drawn as a class diagram and in Figure 6 is shown an example of a curve network drawn as a graph embedded in the plane. A *link* defines the connection between two *nodes* in a curve network and a link can contain many vertices including the nodes. The curve network is given as a list of pointers to nodes. A node is either a start point or an end point of a link. Hence a link has pointers to two nodes. The link between two nodes is again a set of edges (or half-edges in our structure), each of which is given as a line between two vertices. Although it is not necessary, we assume that a node of a curve network also has its relation to a vertex. A node has pointers to a number of links. Note that the geometry of the curve network is represented as points in 2D or 3D given at the vertices.

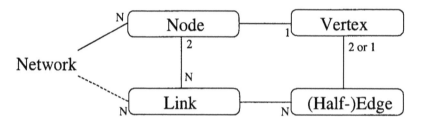

Figure 5. *Standard representation of a curve network.*

Similar to the surface case we can build hierarchies of curve networks. We assume that the topological representation of the curve networks at the different levels are independent, that is, we can retrieve information from any level without using information from other levels.

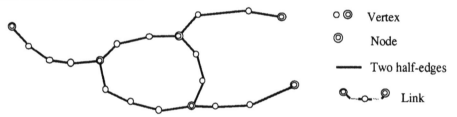

Figure 6. *Curve network as a graph embedded in the plane.*

3.4 Combination

In Figure 7 is shown a complete model where we have joined the curve network structure and the half-edge triangulation structure. The Triangle-Network model (TN-model) consists of one hierarchy of triangle-based surfaces and a set of hierarchies representing various curve networks attached to the triangle-based surfaces. Each curve network consists of a set of nodes. The branches of the TN-model are joined since links are pointing to collections of half-edges. These half-edges are also a part of the triangulation. An example of integrating the terrain and curve network data is shown in Figure 8 and Figure 9.

Figure 8 (a) shows a triangulation of a point set given over a rectangular domain. The black filled circles are vertices and a line between two vertices can be interpreted as a pair of half-edges. Next, we integrate a water curve network with two rivers and a lake boundary, Figure 8 (b). A loop in the curve network is interpreted as a lake. We assume that height values are given at the vertices of the water curve network and that the height values of the water curve network overrule the height values of the triangle-based terrain

surface in Figure 8 (a). The integration goes in three steps:
1. The vertices of the curve network are inserted into the triangulation representing the terrain surface.
2. The edges of the curve network are forced into the triangulation such that each link in the curve network topology becomes a directed graph of half-edges.
3. The terrain is adjusted according to the constraints given by the water curve network, e.g. the height values of points in the lake are adjusted to the altitude of the lake. Note that the points in the lake can be removed from the model. Other requirements imposed by the water curve network are discussed later.

The model as it is shown in Figure 7 now consists of the geometry (the union of the points from the terrain and the points from the water curve network) and the topology given as two branches of the TN-model - a terrain branch and a curve network branch. Note that there is only one level of the hierarchy shown in Figure 8(b). In Figure 9(a) we have integrated a second curve network representing a road. We use the same procedure as in the water case for integrating the road curve network. However, one additional subject has to be taken care of. It is convenient to introduce new nodes where the two curve networks intersect, which in this case naturally can be interpreted as bridges. These nodes have to be inserted into both curve networks. The model now consists of a terrain and two curve networks. We still do not have more than one level of the hierarchy.

In Figure 9 (b) we have removed points from the model shown in Figure 9 (a) without changing the topology of the curve networks. That is, the number of edges (half-edges) and vertices are reduced, while the number of nodes and links are kept the same. The

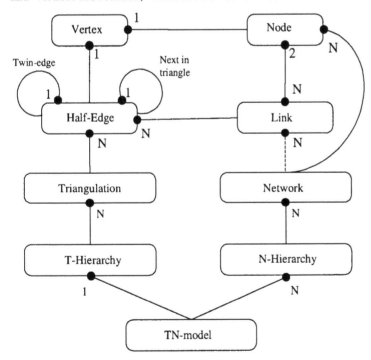

Figure 7. *Integrated topology for terrain and curve networks.*

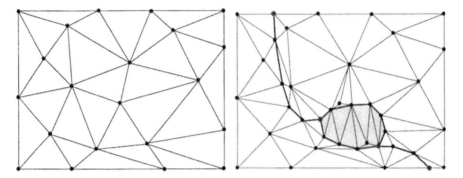

Figure 8. *(a)Triangulation of a point set. (b) A curve network integrated into the Triangulation*

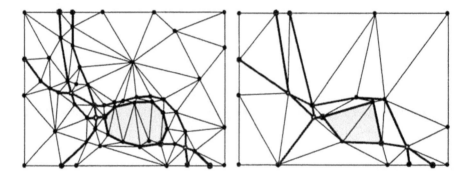

Figure 9. *(a) Two curve networks integrated into the triangulation. (b) Decimated version.*

geometry (points) in Figure 9 (b) is a subset of the geometry (points) in Figure 9 (a). We have now constructed a TN-model with a hierarchy of triangulations representing the terrain at two different resolutions, and two curve network hierarchies each consisting of two curve networks at different resolutions.

3.5 Hierarchy model and construction

A *hierarchical* representation of a surface can be interpreted as a pyramid consisting of a number of layers, where the first layer is the top of the pyramid and the last layer is the base of the pyramid, see Figure 10.

The natural interpretation of this will be that the top of the pyramid (first layer) gives an initial coarse representation of the triangle-based surface, the first together with the second layer a more detailed representation, the first together with the second and third layer an even more detailed representation, etc. Using all the layers of the pyramid gives the most detailed terrain surface.

We assume that the hierarchical structure consists of a number of independent triangle-based surfaces of different levels of resolution. Roughly speaking, we have a coarse-to-fine representation of the triangle-based surfaces from level 0 to level N-1, for some suitable number of levels N in the hierarchy. Since the triangle-based surfaces at each layer are represented independently of each other, extraction or retrieval of a surface at some level n does not need information from the levels i=0,...,n-1. Note that it is only the topological structure that is independent from one level to another. The geometric points behind the vertices are not duplicated. These are stored in one sorted array.

The structure for the hierarchy can be illustrated simply as in Figure 10, where N is the number of layers of the hierarchy.

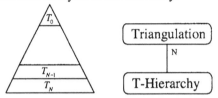

Figure 10. *Hierarchy of triangulations.*

We will here give a short review of the hierarchy construction method used to generate the model and a few examples from a real world data set. We assume that height values are given along the curve networks and that the height values of the curve network overrule the values taken from the scattered terrain data. If values are not given along the curve networks, these values can be found by sampling the terrain. We also assume that the curve network data is consistent with respect to the logic meaning behind the curve network, e.g. that a river system is monotonically decreasing from "mountain" to "ocean". The construction goes as follows:

1. A triangle-based terrain model is constructed from a set of scattered data points using a Delaunay triangulation method [8]. The triangulation is represented as a planar graph.
2. The (half-) edges and the vertices (and nodes) of the curve network is inserted into the triangulation so that the links of the curve network, which is a collection of half-edges, becomes a part of the triangulation, see Figure 9(a). This can be done by constrained triangulation [9]. We denote this model T_n.
3. Given a tolerance, say ε_{n-1}, we remove as many (half-)edges as possible from the model T_n so that the difference between the new model T_{n-1} and T_n is less than ε_{n-1} in some appropriate norm. This can be done in various ways, see Heckbert and Garland (1997). Alternatively, we can choose to remove a certain percentage of the edges. The removal has to be done in such a way that the topology of the curve network is maintained. Again this is accomplished by using constrained triangulation or by collapsing edges of the triangulation as it is done here. The edge collapse technique is presented in Daehlen and Fimland (1999). The new model consists of both T_n and T_{n-1}. That is, two terrain surfaces at different resolution and a set of curve networks, each curve network represented at two resolutions.
4. Finally, we repeat the process for tolerances ε_{n-2},..., ε_0 in order to obtain the complete model T_n, T_{n-1},...,T_0.

3.6 Example

A variety of methods can be used for selecting data at the different levels of the hierarchy, see Heckbert and Garland (1997). In this section we briefly go through an example, based on the implementation presented in Fimland and Skogan (2000). The data model and implementation presented in this paper form a basis for the data selection and construction method presented in Fimland and Skogan (2000).

The illustrations in the Figures 11-14 show some results where the data model and the hierarchy construction method are used. The terrain is constructed from approximately 64.000 points (and therefore we obtain about 128.000 triangles from the initial Delaunay triangulation). Curve network data from roads, rivers (small and large) and lakes are combined and integrated into the terrain model using a constrained Delaunay triangulation procedure. The total number of geometric edges in the combined curve network is approximately 1200. Figure 11 is a picture of the whole data set.

In the Figures 12-14 we show pictures of parts of the whole data set. Figure (a) shows an area around a lake at the finest level (high resolution) and Figure (b) shows the same at a coarser level (lower resolution). The coarse representation in Figure (b) uses approximately 25% of the original data used in Figure (a). As can be seen, some of the details have been removed (e.g. the shape of the island in the lake), while the main structures can easily be recognised from the original data set. Most important however, the topology of the curve network is unchanged and no new intersections between the curve network and the terrain models have been created. Thus, the decimation process has ensured *consistency* in the data sets.

Figure 11. *Overview of the data set at the finest level*

In Figure 13, the same principles are illustrated. A river runs through the bottom of a valley while several smaller rivers run into the main river. The road curve network (drawn in black) crosses the river curve network at multiple places. At the coarser level, the main contours of the terrain are still present, while the details vanish. Figure 14 shows the same data sets using wire-frame visualisation. Again we observe that the intersections between the terrain and the curve networks and in-between the curve networks are unchanged. Thus, the topology is maintained.

4   Operations

There are many issues to be studied in connection with the work presented in the previous sections. We will here give a brief discussion on the operational aspects of the framework. Important operations are data pre-processing, construction, integration, consistency control, update, visualisation and customisation.

Figure 15 shows the main parts of the high level functionality of the framework. Working with huge data sets over huge regions will in general require some pre-

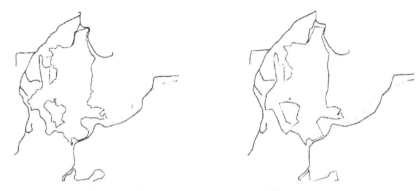

Figure 12. *A lake at the finest level (a) and a coarser level (b)*

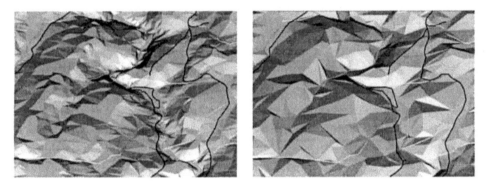

Figure 13. *A small lake and a river at the finest level (a) and a coarser level (b)*

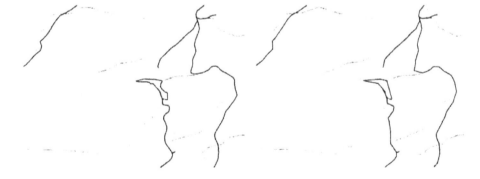

Figure 14. *A small lake and a river at the finest level (a) and a coarser level (b) shown using wire-frame visualization*

processing before constructing terrain surface objects and their attributes. If the underlying domain were divided into subdomains or tiles, we would like to partition the data into sets associated to each tile. In the figure this is referred to as segmentation. The critical challenge, however, becomes the task of ensuring proper behaviour on the borders between the tiles. For visualisation purposes this problem is studied in Sevaldrud (1999). It is basically a question of maintaining the continuity between the tiles. We want surfaces and curves defined on different tiles to join "smoothly" across tile's boundaries.

Moreover, it might be convenient to sort the points based on some suitable criteria, e.g. density. In many cases it is also important to verify that the data fulfil certain properties, for example, there might be outliers or some other systematic error in the data.

The generation of hierarchical surfaces is basically interpolation and approximation of scattered data where the final model is represented as a hierarchical triangular model. It is important to distinguish between construction of a complete model and the updating of a model based on limited input. Correction of the surface, both geometrically and topologically, is necessary when introducing new information. Once a complete data set has been collected and the initial build-up of curve network integration and terrain modelling has been performed, the need for updating will arise. A complete rebuild for every change in the data set will be time-consuming. Instead we will use a strategy which updates only a portion of the hierarchical structure. A correction of the data set will have consequences in all levels of the hierarchy. Furthermore, a correction can be made at any level and an update procedure will have to ensure that changes are reflected throughout all other levels of the hierarchy. This provides an important extension of the work presented here, however, further investigations are necessary. The domain decomposition and the updating procedures are indicated as a part of the representation and functionality of the software, respectively. This is indicated in Figure 15.

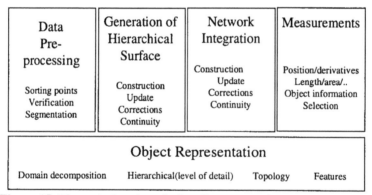

Figure 15. *The main functions for handling triangulations and curve networks integrated into the triangulations.*

Finally, integration of curve network data into the surface model (or carrier) involves the same aspects as the construction of triangle-based surfaces. And with respect to the surface construction it involves imposing edge constraints into the triangulation. An important aspect for further investigations is how these constraints shall be imposed into a hierarchical structure.

5   Summary And Future Work

In this paper we have presented a framework of three major parts. First part is the context of the geodata pipeline and the geodata carrier where curve network and terrain data are represented in a triangulated irregular curve network. The second part is the data model, which in detail describes how the data is logically organised into a multi-resolution data structure, according to a hierarchical represented layered model. The data model allows us to store a consistent integrated model at different levels of resolution. The hierarchical layered model gives a compact representation as well as efficient access to the data. The third part is the operational aspects of the framework. We have described a set of operations that interacts with the data model, which may be considered an integral part of the data model in an object-oriented implementation.

The implementation described in Fimland and Skogan (2000), shows that the framework is implementable and leads to more consistent data maintenance and efficient access. However, it has potential for improvements. Especially the data model should be evaluated and inter-resolution dependencies should be taken into account to ensure incremental updates. Other issues that should be addressed are the ability to handle large data sets and better support to resolve inconsistencies.

Acknowledgement

This work has been supported by the Norwegian Research Council's DYNAMAP-I project 118048/223. The authors would like to thank David Skogan for helpful suggestions and comments.

References

Heckbert, P.S. and M. Garland, *Survey on Polygonal Surface Simplification Algorithms*, 1997, Multiresolution Surface Modeling Course (SIGGRAPH'97).

Cignoni, P., E. Puppo, and R. Scopigno. *Representation and visualization of terrain surfaces at variable resolution*. in *Scientific Visualization '95*. 1995: World, Scientific.

De Floriani, L. and E. Puppo, *Hierarchical triangulation for multiresolution surface description*. ACM Transactions on Graphics, 1995. 14(4): p. 363-411.

Jones, C.B., D.B. Kidner, and J.M. Ware, *The Implicit Triangulated Irregular network and Multiscale Spatial Databases*. Computer Journal, 1994. 37(1): p. 43-57.

Ware, J.M. and C.B. Jones, *A Multiresolution Topographic Surface Database*. International Journal of Geographic Information Systems, 1992. 6(6): p. 479-496.

Dæhlen, M. and M. Fimland. *Constructing Hierarchical Terrain Models by Edge Ordering over Triangulations*. in *7th Scandinavian Research Conference on Geographical Information Science (ScanGIS '99)*. 1999. Aalborg, Denmark.

Sevaldrud, T.E., *Hierarchical Terrain Models with Applications in Flight Simulation*, in *Department of Informatics*. Master Thesis, 1999, University of Oslo.

Nielson, G.M., *Tools for triangulation and tetrahedration and constructing functions defined over them*. IEEE Computer Society Press, 1998. Scientific Visualization: Overviews-Methodologies-Techniques: p. 429-525.

De Floriani, L. and E. Puppo, *An On-line Algorithm for Constrained Delaunay Triangulation*. CVGIP: Graphical Models and Image Processing, 1992. 54(4): p. 290-300.

SINTEF, *The SINTEF Scattered Data Library (Siscat)*, http://www.oslo.sintef.no/siscat, 1995, SINTEF Applied Mathematics.

Weiler, K., *Edge based data structures for solid modeling in curved-surface environments*. IEEE Computer Graphics and Applications, 1985. 5(1): p. 21-40.

Fimland, M. and D. Skogan. *A Multi-Resolution Approach for Simplification of an Integrated Network and Terrain Model*. in *9th International Symposium on Spatial Data Handling*. 2000. Beijing, China.

Wilson, R.J., *Introduction to Graph Theory*. 1985, Essex, UK: Longman.

Behzad, G.C.M. and L. Lesniak-Foster, *Graphs and Digraphs*. 1979, Belmond, CA.: Wadsworth.

Mchugh, J.A., *Algorithmic Graph Theory*. 1990: Prentice-Hall Inc.

LEDA, *A library of the data types and algorithms of combinatorial computing*, 1988.

7

Integrated Spatiotemporal Analysis for Environmental Applications

Robert Frank and Zarine Kemp

1   Introduction

Geographic information systems (GIS) have been used successfully in many applications where digital representations of spatial resources and their management is a prime requirement. However, when using GIS for environmental management and decision support, many more considerations come into play. This awareness of the need to extend basic GIS functionality to encompass spatiotemporal data management as well as computational modelling capabilities has been documented (Abel 1993, Claramunt 1998). A number of current proprietary GIS are characterised by their flexibility and can be used for certain environmental applications but do not provide the specialised functionality required in the environmental GIS (EGIS) domain. Likewise, custom-built applications are good at solving specific problems but have very little generic flexibility. Both approaches often lack the facilities, extensibility and functionality that are required for environmental research which can be characterised by:

- large volumes of data, collected using a multiplicity of sensors which need to be processed before being used for analysis; the problem is often described as one of 'too much data but not enough information'
- the need to merge raster, vector, scalar, hierarchical data with researchers' inherent domain knowledge
- requirement to access, and make available, data over local and wide-area networks
- the need for open software architecture that is extensible and flexible.

This paper proposes a component-based framework which overcomes these limitations and is flexible enough to respond to the support needed for an EGIS. The framework presented makes it suitable, not just for the marine biodiversity example application discussed in this paper, but also for a wide range of environmental spatiotemporal research. The paper is organised as follows. Section 2 presents a motivating example to illustrate a typical environmental modelling application. Section 3 uses the example for a rigorous analysis of generic EGIS requirements. Section 4 discusses the design principles underlying the framework. Section 5 presents the prototype system design and section 6 concludes the paper.

2   Motivating Example

The Dover Straits and the Southern North Sea, within the coordinates of 1'00E 51'00N / 3'00E 52'00N is well known as one of the busiest shipping routes in the world, but also houses a complex marine habitat. In collaboration with other institutions, an ongoing project (Cottonec *et al.* 1998) aims to develop an environmental model of the biodiversity of the area, and build a general data management and analysis system which will support researchers in exploring and gaining insights into marine environments. Observational data sets captured during research cruises consist of biological data for plankton and other species identified by species and stage of growth if possible, or if that is not feasible, grouped at a higher taxonomic level. The taxonomic hierarchy, with its implicit classification 'inheritance' relationships between the taxa forms part of the underlying knowledge base for the marine application domain. Water current and trawl duration data are also captured as they are relevant parameters for calculating species abundance values. Observational data capture also includes sensor-derived, georeferenced physical oceanic attributes such as temperature, salinity and density.

The sampled data sets are combined with data from other sources to enable a realistic representation of the problem space within the system. For example coastline and bathymetry data in vector format are used to spatially delineate the ecosystem. Depending on the requirements of a particular problem, the land sea boundary may need to be specified with reference to tidal models. Other oceanographic variables that are remotely sensed such as sea surface infra-red images are in raster format and macro-scale hydrographic data generated by ocean models are in gridded format. In order to integrate these diverse data sets the environmental modelling system must provide the computational tools to ingest and transform the data to enable multivariate analyses (Wright *et al* 1997, Wright and Goodchild 1997). Figure 1 illustrates the data capture mechanisms that are involved.

Experience with handling the data in the project has elicited some surprising conclusions. Although environmental systems are often overwhelmed by data which can easily go over the gigabyte mark, necessary information for building a consistent model is often missing. Just one trawl with the CPR (continuous plankton recorder) or other automated robots can collect a high density of information, but usually supplying just one data attribute, e.g. plankton abundance. The closest temperature measurement in that area may be two days old and other data may not even exist. Even our own sampling, with respect to the most detailed physical, chemical and biological parameters, along the coastlines took several days and therefore the raw data cannot be treated as a consistent snapshots of a specific region of the marine environment. To enable the system to deal with such erratic dispositions of data in time and space, geostatistical methods are required to investigate the data, assessing their usability in a 4-D model and for calculating a valid spatiotemporal range.

3   Analysis Of EGIS Requirements

From the description of the marine environment application in the previous section it is evident that the requirements for a supportive system can be considered in terms of: the

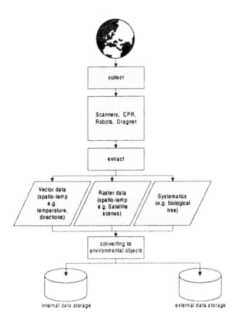

Figure 1. *Input sources for environmental data*

data capture process, transformation of the data into internal representations that can be used for exploratory analysis and the provision of abstractions to enable system level information to be mapped into concepts relevant to environment modellers. A crucial component of any GIS is, of course, the visualisation module. We assume that visualisation functionality is provided as a component of the interface as indicated in sections 4 and 5. However, we do not discuss visualisation requirements here as they are not unique to environmental monitoring applications. In this section we concentrate on the special data structures and processing functionality that are fundamental to an EGIS.

3.1 Data capture

Within most EGIS exploratory data analysis is basic requirement. Standard statistical capabilities are needed to enable the researcher to explore the relationships underlying the variables as well as to investigate the range and distribution of individual attributes. This is partly a consequence of the relative sparsity of data relevant to the marine environment and partly the uneven spatial coverage.

One of the deficiencies of most proprietary GIS is that they do not support the user in determining 'fitness-for-use' measures for data sets. They read, store and display the data without providing capabilities for the researcher to evaluate data quality. Usually the data comes from a wide variety of sources, and formats. Sometimes information about the sampling techniques and lineage is not available (von Meyer 2000). These uneven sampling strategies cause large data clusters in space, time and quality, and the data-clustering is exaggerated due to the fact that robots can transmit large volumes of data within a short timeframe thereby creating undesirable skewness. The only spatially uniform distributed data are the satellite images that are treated as snapshots after pre-processing even though they were originally captured by pushbroom-scanners. Though satellite images are not perfect (when generating interpolated data close to coastlines the resolution can be too coarse) an aim is to use them as a benchmark to validate related data.

As an example of preliminary data exploration and evaluation, consider the relationship between temperature, salinity and density as illustrated in Tables 1 and 2 and Figure 2. Salinity is positively correlated with density, while both have a negative correlation to temperature as shown in Table 1. These dependencies are further confirmed by extracting factors to describe the composite data set. The results of the factor analysis are presented in Table 2 and 3 and Figure 2. Figure 2 illustrates the plot of the eigenvalues on the left hand side and a 2d factor plot on the right hand side for temperature, salinity, density fluorescence, irradiance and transmission. Two factors with an eigenvalue of more than 1 could be extracted. They explain the dataset by 78.54 %. The first component mainly is driven by the physical parameters temperature, salinity, density and transmission, while the second component is determined by the biological factors fluorescence and irradiance. Hence there is a clear discrimination of a physical (*Comp. 1*) and a biological factor (*Comp. 2*) in the data set (Table 3).

	TEMP	SAL	DENS	FLUO	IRRAD	TRANS
TEMP	1.000	-0.527	-0.690	0.637	-0.071	0.463
SAL	-0.527	1.000	0.979	-0.303	-0.098	-0.672
DENS	-0.690	0.979	1.000	-0.411	-0.066	-0.684
FLUO	0.637	-0.303	-0.411	1.000	-0.323	0.161
IRRAD	-0.071	-0.098	-0.066	-0.323	1.000	0.239
TRANS	0.463	-0.672	-0.684	0.161	0.239	1.000

Table 1. *Correlation matrix of temperature, salinity, density, fluorescence, irradiance and transmission on sea surface level*

Component	Eigenvalue	Pct of Var	Cum Pct
1	3.284	54.726	54.726
2	1.429	23.815	78.54
3	0.627	10.449	88.99
4	0.379	6.312	95.302
5	0.282	4.697	100
6	2.403E-05	4.005E-04	100

Table 2. *Computed eigenvalues and percentages of the marine dataset*

	Comp. 1	Comp. 2
TEMP_S	**0.701**	0.497
SAL_S	**-0.921**	-4.130E-02
DENS_S	**-0.954**	-0.155
FLUO_S	0.367	**0.796**
TRANS_S	**0.836**	-0.171
IRRAD_S	0.261	**-0.791**

Table 3. *Rotated component matrix*

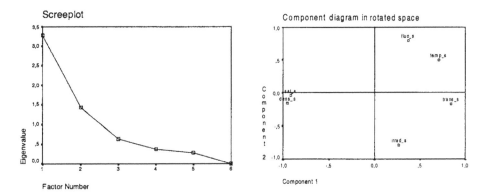

Figure 2. *Graphical representation of the extracted factors*

3.2 Selection and subsetting of data

Although data in a typical EGIS are heterogeneous and multisource, they invariably have spatial, temporal and thematic dimensions associated with them. The data can, conceptually, be thought of as a multidimensional hypercube (Lee and Kemp 2000). Depending on the particular analytical problem being solved, only a partial subset of this hypercube is required at any one time. Therefore, an on-the-fly, flexible subsetting capability is also required in an EGIS to enable 'slicing' and 'dicing' through all dimensions of the data hypercube. For example, <select *"Calanus finmarchicus"*, from x_1, y_1, x_2, y_2, for $t_1..t_2$> would extract all abundance values for the required plankton species that occurred in the area specified by the bounding box x_1, y_1, x_2, y_2 for the temporal interval defined by $t_1..t_2$. Alternatively, the researcher may wish to extract accumulated abundance values for all species within a given genus or family. In this case the taxonomic hierarchy, which is represented in the data store as a domain-dependent knowledge object is invoked by the system to determine which species values are to be aggregated. This data selection capability is also required to return consolidated values such as averages, counts and other measures that are referred to as aggregate functions in standard database query languages. Flexible data extraction is easier to achieve if the EGIS is implemented on top of a relational or, better still, object-oriented database management system with its query and retrieval facilities made available to the researcher at the user interface.

3.3 Representation of Data

In multivariate analyses, in order to perform any meaningful spatial operations it is necessary to transform attributes into comparable structures (Mason *et al.* 1994, Lucas 2000). Environmental modelling deals with attributes that are by and large continuous over the problem space so the most obvious data structures to use are 2-, 3- or 4-d grids or meshes (Kemp and Lee 1998, Musick and Critchlow 1999). Conversion of point sampled observations to grids requires geostatistical capabilities to enable data structures of varying resolution to be generated. As we explain below, in some situations, standard interpolation techniques may be inadequate so tailor-made procedures may be have to be included in the function library.

An understanding of how the data values vary as a function of distance allows interpolation of values at unsampled locations. The basic assumption is that the relationship between the value of a point and any sample value depends on the distance and possibly on the direction. These techniques rely on an understanding of the spatial correlation of the structure of the data that is used to guide the interpolation (Burrough and McDonnell 1998). An isotropic spatial correlation is dependent on the distance, while the much more common anisotropic correlation is dependent on direction and distance. The two minimum conditions, which are sufficient for geostatistical analysis, are stationarity and spatial correlation (Legendre and Trousellier 1988). Therefore it is important to test these properties in the data set so that geostatistical assumptions can be made and correct weighting factors selected to represent the marine channel environment.

Most interpolation methods, including nearest neighbour and inverse distance weighting assume a linear or factor weighted spatial correlation between all sample points. In contrast to reality, this is a simplification, because there are several circumstances that may affect a uniform distribution like the shape of a coastline, water currents or biological interactions. In a marine environment, it is reasonable to assume that there is a higher spatial correlation along the coastline than vertical to the coastline as we usually expect a more even distribution along a coastline than vertical to it. These anisotropies represented as ellipsoids should be calculated for estimating a range for the reliability of the data. The search-ellipses indicate the statistical and average range of a specific value in two, or as an ellipsoid in three dimensions (Grioche *et al.* 1996). There are three ways of enhancing the geostatistical results: stretching the coastline to a straight line, partitioning the data set depending on the orientation of the coastline and variogram modelling in multiple directions while changing the lag spacing. All three methods have inherent problems but variogram modelling in multiple directions and changing the lag spacing, although it will not give the complete range, avoids geometric problems.

The key to a successful marine model is the recognition and handling of the determining parameters. The physical values as shown below are indicators for the biodiversity in the Dover straits. Even though their influence and interactivity varies within different biological life stages they play an important role in all development stages. Within a life cycle, phytoplankton may accumulate at an early stage in low salinity and cold water but at a later stage it is most often sampled in low salinity but very high temperature regions. Figures 3 and 4 illustrate examples of retrieval of subsets of data and visualisation of the resultant 2-d grid structures. The research analysis involves considering the effects of the variables density, temperature, fluorescence and salinity on the growth of fish larvae in the region during spring. Figure 3 presents the physical variables at 3 depths through the water column, surface, bottom and mid-range. Figure 4 shows abundance values of larvae through three stages of growth.

3.4 Derived Environmental Objects

From a biological point of view a marine environment is a collection of biotopes, while from the data modelling perspective, these biotopes are compositions of specific physical, chemical and biological factors. In contrast to most other environments the marine ecosystem is very dynamic. It is influenced by temporal effects that range from seasonal changes to short-term tidal influences. Furthermore, the biological factors that regulate biological rates are non-linear and make interpolations difficult when point samples are taken during a period of time (Brandt *et al.* 1996). The definition of a marine ecotope can become fuzzy when it is just based on biological data. During the genesis from ichthyoplankton to fish, not only do the organisms prefer different marine habitats, but it is also difficult to determine discrete boundaries. As each fish species is partly a product of regional oceanography, coastal geomorphology, habitat availability and natural disturbance the habitat can be defined by more steady physical parameters, e.g. combining the bottom type with the water-depth (Schmidt *et al.* 1999). This fuzzy information can be stored as environmental data objects where often changing threshold-values can easily be modified without redesigning the whole data model. An EGIS should enable users to define these domain relevant objects as part of the knowledge base.

There are various techniques for representing this type of derived data within the EGIS (Breunig 1999). A compositional object specification may be used, defining object structure and behaviour to represent environmental objects. Alternatively a rule-based expert system, embedded in the EGIS can be used to represent the way in which a biotope may be instantiated from the underlying data sets.

3.5 Requirements summary

To meet the wide range of requirements discussed, the following criteria were identified for an environmental GIS:

Structural:
- management of heterogeneous spatiotemporal data at multiple scales and in different formats; this implies the processing required to transform data to required internal representations and classify observed data
- the data manager module must be extensible to enable incorporation of new variables pertinent to the research; the data servers may be internal or remote from the EGIS framework
- the EGIS must enable derivation of data at different abstraction levels; depending on the requirements of the analysis, 'natural units' may vary

Procedural:
- provide a spatial statistical toolkit; for exploratory analysis of data sets, for surface generation, error analysis and hypothesis testing
- enable process models, to be invoked by the user at various scales, either to derive predictive data through time or to determine spatial dispositions of species that may be linked via predator-prey relationships
- graphics and visualisation capabilities; in a computational environment where the spatial properties predominate; 2d and 3d cartographic visualisation is a major requirement both as an aid to exploratory analysis as well as for presentation of multidimensional interactions

4   Design And System Framework

4.1   Overview of architectures for environmental information systems

With increasing interest in solving environmental problems and growing world wide effort in 'digital earth' projects, there is a wide spectrum of software architectures used to implement these systems. Many excellent systems have been designed to target specific problems, to enable decision support for particular organisations or to solve particular

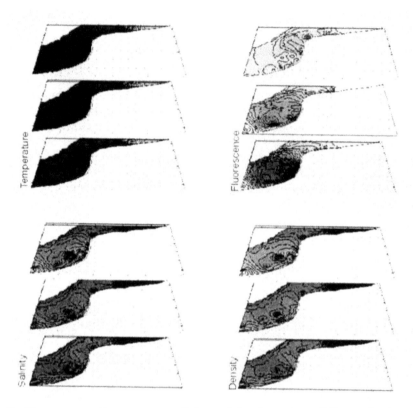

Figure 3. Temperature, fluorescence, salinity and density at three different depths

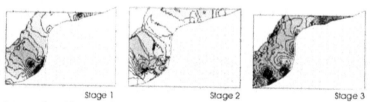

Stage 1   Stage 2   Stage 3

Figure 4. *Interpolated larvae stages*

environmental problems using predefined data sets (Goodchild *et al.* 1996). Several of these systems have been built using proprietary GIS products that have become increasingly sophisticated in the functionality and the interfaces that they provide. However, for a flexible but supportive EGIS we contend that using a single proprietary system is not adequate because:

- most GIS are monolithic systems so flexible data management is not within the developer's control; several GIS now enable links with general-purpose DBMS systems to overcome this problem.
- the use of a proprietary product generally constrains the developer to a user interface hosted within the product. In an EGIS, the user interface is a critical component of the system and should be within the developer's control and the proprietary system

may be used as a sub-component. Some products such as ESRI's MapObjects (discussed in the next section) do provide this capability.

- in an EGIS, time is a critical dimension as most analyses involve representation of time-varying objects and evaluation of changes through time (Panzeri and Morris 2000). The data structures provided by most GIS do not provide much support for representation and management of time.

- in the marine EGIS domain, the management of 3-dimensional data structures (4-d if time is included) is not handled very well by third party systems. A generic EGIS can provide this capability as a core function.

Given these constraints we considered two options for the marine EGIS: using an object-oriented database management system (OODBMS) and using a modular, component-based software engineering approach.

4.2 Object-oriented approach

Much of the core functionality of a marine EGIS can be provided by building it on top of an object-oriented data server (Bernard et al. 1998, Patel *et al.* 1997). An OODBMS not only provides capabilities for modelling and management of the data, but also provides an interface to external modules and libraries. Environmental data like satellite images, current directions and speeds or plankton types and quantity per volume cannot be easily represented in a relational database. The complex data-structuring capabilities within OODBMS provide a more natural mapping between environmental and spatial objects and the underlying software (Blaha and Premerlani 1998, Clementini and Felice 1994, Davis and Albuquerque de Vasconcelos Borges 1994). This strengthens DBMS support for flexible modelling as the physical data implementation is closer to the logical design and leads to more compact storage. Another reason for considering an object-oriented modelling and implementation substrate for EGIS is that it enables the encapsulation of the structural and procedural requirements that are closely inter-related. As far as possible the spatial and temporal data objects conform to internationally accepted standards such as those proposed by the OpenGIS Consortium (Schell 1995) or the Canadian Spatial Archive and Interchange Format (SAIF). An extensible object-oriented model enables the base object types to be extended or aggregated as required. For example, the definition of an *estuary* biotope would require specification of an aggregation object that is specified by a collection of references to related objects. We have also identified the problem of querying across multi-scale multi-format data. The requirement here is to enable seamless transformation to achieve consistent scales for queries. These transformation functions can be embedded as *methods* in individual object types or via a library of standalone transformation functions.

The provision of software libraries for statistical and mathematical operations is more problematic. A dedicated library can be built into the EGIS and its computation model made available in the user interface. However, it is less easy to incorporate data and services available over the Internet. In the latter case, middleware would be required to interface between the systems. With respect to interoperability between distributed data sets, the concept of virtual data sets (VDS) similar to OGIS can be used (Vckovski 1998). The data exchange is not based on a standardised data structure, but by a set of interfaces. This easily allows expanding the database over a network since the data objects are not accessed from a physical file, but are called by a set of corresponding methods in real-time.

4.3   Component-based approach

Component-based software development is becoming popular partly due to support from industry leaders (Chappell 1996, Otte 1996) and partly because it addresses the need for extensible, flexible software engineering solutions. The infrastructure enables the system to provide transparency with respect to distribution and hardware/software platforms. When working with software components, generic procedures can be modularised, and hence can become more independent with an open data interface. This interface enables data and procedure exchange that may thus be reused by several applications. Earlier discussion has identified the need for special functionality for an EGIS (geostatistics, support for fuzzy data); these can be incorporated in an extensible manner if the structure of the EGIS consists of independent modules in a way that they easily can be exchanged, added and altered. The more autonomous these components are, the more easily they can be used for a specific customised system. In the design for the marine EGIS, autonomous components were identified to:

*   manage the heterogeneous data sources needed; the objective here is that each source can be managed by a particular server which deals with the unique type and format of data expected
*   provide a set of procedure libraries; the paper has identified requirements for standard statistical tools, specialised geostatistics, a 2-d and 3-d visualisation and mapping toolkit, graphing and report facilities etc.
*   user interface management; an open ended modular system requires an intelligent interface that can guide the user towards appropriate functions.

5   The Prototype System

After consideration of the pros and cons of different development environments, it was decided to develop the prototype marine EGIS system initially using a component based approach as described in section 4.3. The main system components are illustrated in Figure 5.

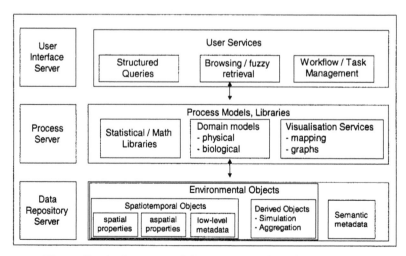

Figure 5. *Architecture of the component-based EGIS*

The framework is implemented in Microsoft Visual Basic (Pattison 1998) and uses ESRI's MapObjects (Hartman 1997) for mapping and cartographic functionality. There are several reasons for choosing this software substrate. The user interface in a system such as an EGIS is a major factor in its acceptance by the research community. It was important that the functionality be presented in the manner and sequence in which the users approach the exploratory tasks. Control of the structure and look-and-feel of the interface was easier to achieve using the component approach.

Another consideration was the fact that it was more logical to use generic components that could be reused. In a component-based environment the modules can be specially written or existing third-part components can be plugged in. We use a mixture of the two strategies; MapObjects provides a basic set of mapping and GIS components similar to the functionality of ArcView which makes EGIS development easier and enables reuse of these components to customise similar systems. However, we have been constrained by the fact that the version available did not extend to the spatial analysis capabilities for manipulating raster or grid types so some of the components were custom written. A forthcoming version of the software will make the task of implementing the GIS more straightforward.

The component model has also enabled us to manage the underlying database independently of the EGIS computational engine. This makes the data subset selection tasks a lot easier to implement. Currently Microsoft Access is used to store the data but that is not an ideal solution as some object database capabilities are thus not available. In later versions of the marine EGIS we intend to make the design even more modular by separating the data servers from the main system. This will enable the data management facilities to be configured separately for a range of data servers. This is particularly important in an area such as environmental modelling and analysis where research communities are very dependent on using and sharing data globally. Each server will be designed to manage a data set of a particular type and format and deal with local as well as globally distributed data repositories (Fox and Bobbitt 2000).

6 Conclusions

The results so far have been promising and prove that the basic requirements identified for a supportive exploratory research framework are essential. It is important to note that the rationale underlying the framework was not to provide the user with a predetermined set of functions that could be selected to invoke required tasks in an automatic manner.

The environmental modelling community consists of users who are expert in their respective fields with an understanding of the hypotheses they wish to explore and the tools that they want to use. In other words, they wish to have control over the workflow in the analytical process. Consequently, the core purpose of the design of the generic EGIS framework was to make generic data structures and procedural libraries available in a convenient and easy to use manner. The researcher is insulated from the inconvenience of switching between custom-built packages and their tailored interaction mechanisms but does not lose any of the flexibility provided by a rich set of toolkits necessary to solve complex environmental interactions and visualise change. The ongoing research includes some of the future work identified in section 5, above.

Acknowledgements

The first author Robert Frank is supported by INTERREG Grant KC 97/05 Biodiversity and Cartography of Marine Resources in the Dover Strait (GOSE reference 97/C8/03). We gratefully acknowledge the contribution of the other members of the project at the Durrell Institute of Conservation and Ecology, University of Kent, Canterbury Christ Church University College and the Marine Laboratory, Université du Littoral, Calais.

References

Abel, D. and Ooi, B.C. (1993) Advances in Spatial Databases, in: D. Abel and B.C. Ooi (Eds.) *3rd International Symposium, SSD '93, June 23-25 1993*, 529 p., Springer Verlag, Singapore, ISBN 0387568697.

Bernard, L., Schmidt, B. and Streit, U. (1998). AtmoGIS - Integration of Atmospheric Models and GIS, in: T.K. Poiker and N. Chrisman (Eds.), *Proceedings of 8th International Symposium on Spatial Data Handling*, July 11-15 1998, 267-276 p., International Geographical Union, Vancouver.

Blaha, M. and Premerlani, W. (1998) *Object-Oriented Modelling and Design for Database Applications*, 484 p., Prentice-Hall, Inc., New Jersey, ISBN 0-13-123829-9.

Brandt, S.B., Mason, D.M., Goyke, A., Hartmann, K.J., Kirsch, J.M. and Luo, J. (1996) Spatial Modelling of Fish Growth: Underwater Views of the Aquatic Habitat, in: M. F. Goodchild, L. T. Steyaert, B. O. Parks, C. Johnston, D. Maidment, M. Crane and S. Glendinning, (Eds.), *GIS and Environmental Modeling: Progress and Research Issues*, 225-229 p., Parker, Dennisos H. Sherwood, Nora, Fort Collins, USA, ISBN 0470236779.

Breunig, M. (1999) An Approach to the Integration of Spatial Data and Systems for a 3D Geo-Information System, *Computers and Geosciences*, 25, 39-48 p.

Burrough, P. and McDonnell, R.A. (1998) *Principles of Geographical Information Systems*, 2nd edition, Oxford University Press, ISBN 0198233655.

Chappell, D. (1996) *Understanding ActiveX and OLE*, 328 p., Microsoft Press, Redmond, Washington, ISBN 1572312165.

Claramunt, C., Parent, C., Spaccapietra, S. and Theriaut, M. (1998) Database Modelling for Environmental and Land Use Changes, *Geographical Information and Planning: European Perspectives*.

Clementini, E. and Felice, P. D. (1994) Object Oriented Modelling of Geographic Data. *Journal of the American Society for Information Sciences* 45, 694-704 p.

Cotonnec, G., Grioche, A. and Harlay, X. (1998) Biodiversity and Cartography of Marine Resources on both Sides of the Straits of Dover, in: P. Koubbi, B. Sautour and M. Walkey (Eds.) 20 p., *Interreg II News*.

Davis Jr., C.A. and Albuquerque de Vasconcelos Borges, K. (1994) Object-Oriented GIS in Practice, *URISA 1994*.

Fox, C.G. and Bobbitt, A.M. (2000) The National Oceanic and Atmospheric Administration Vents Program GIS: Integration, Analysis and Distribution of Multidisciplinary Oceanographic Data, *Marine and Coastal Geographical Information Systems*, D. Wright, D. Bartlett (Eds.), Taylor and Francis, London, UK, ISBN 0-7484-0862-2.

Goodchild, M.F., Steyaert, L.T., Parks, B.O., Johnston, C., Maidment, D., Crane, M. and Glendinning, S. (Eds.) (1996) *GIS and Environmental Modeling: Progress and Research Issues*, GIS World Books, Fort Collins, CO, USA, ISBN 0470236779

Grioche, A., Koubbi, P., and Harlay, X. (1996) Spatial Patterns of Ichthyoplankton Assemblages along the Eastern English Channel French Coast during Spring 1995, *Proceedings of 20th Larval Fish Conference,* Jun 13-19 1996, 14 p., Fisheries Institute, Louisiana State University, Baton Rouge LA, USA.

Hartman, R., (1997) *Focus on GIS Component Software featuring ESRI's MapObjects*, 400 p., Onward Press, Santa Fe, USA, ISBN 1-56690-136-7.

Kemp, Z., and Lee, T.K.H. (1998) A Marine Environmental Information System for Spatiotemporal Analysis, in: T.K. Poiker and N. Chrisman (Eds.), *Proceedings of 8th International Symposium on Spatial Data Handling*, July 11-15 1998, 1998, 474-483 p., International Geographical Union, Vancouver.

Lee, T.K.H. and Kemp, Z. (2000) Hierarchical Reasoning and on-Line Analytical Processing in Spatiotemporal Information Systems, in: P. Forer, A.G.O. Yeh, J. He (Eds.), *Proceedings of SDH 2000, 9th International Symposium on Spatial Data Handling*, August 10-12 2000, 3a.28-40 p., Beijing, P.R. China.

Legendre, P. and Troussellier, M. (1988) Aquatic Heterotrophic Bacteria. Modelling in the Presence of Spatial Autocorrelation, *American Society of Limnology and Oceanography,* 33, 1055-1067 p.

Lucas, A. (2000) Representation of Variability in Marine Environmental Data, *Marine and Coastal Geographical Information Systems*, D. Wright, D. Bartlett (Eds.), Taylor and Francis, London, UK, ISBN 0-7484-0862-2.

Mason, D. C., M. A. O'Conaill, Bell, S.B.M. (1994) Handling Four-Dimensional Geo-Referenced Data in Environmetal GIS. *International Journal of Geographical Information Systems* 8(2): 191-215 p.

von Meyer, N., Foote, K.E. and Huebner, D.J. (2000) Information Quality Considerations for Coastal Data, *Marine and Coastal Geographical Information Systems*, D. Wright, D. Bartlett (Eds.), Taylor and Francis, London, UK, ISBN: 0-7484-0862-2.

Musick, R. and Critchlow, T. (1999) Practical Lessons in Supporting Large-Scale Computational Science, *SIGMOD Record*, 28(4), December 1999.

Otte, R., Patrick, P., Roy, M. (1995) *Understanding CORBA: The Common Object Request Broker Architecture*, 288 p. Prentice Hall, Upper Saddle River, N.J., USA, ISBN 0134598849.

Panzeri, M. and Morris, K. (2000) The Integration of Spatial and Temporal Data for Consortia Based Initiatives: The use of a 4D GIS, *Proceedings of Oceanology International 2000*, 337-347 p., March 7-10 2000, Brighton, UK.

Patel, J., Yu, J., Kabra, N., Tufte, K., Nag, B., Burger, J., Hall, N., Ramasamy, K., Lueder, R., Ellmann, C., Kupsch, J., Guo, S., Larson, J., DeWitt, D. and Naughton, J. (1997) Building a Scalable Geo-Spatial DBMS: Technology, Implementation, and Evaluation, *Proceedings of SIGMOD 1997*, May 13-15 1997, 336-347 p., Tucson, AZ, USA, ISBN: 0897919114.

Pattison, T. (1998) *Programming Distributed Applications with COM and Microsoft Visual Basic*, 300 p., Microsoft Press International, Redmond, Washington, ISBN 1572319615.

Schell, D. (1995) What is the Meaning of Standards Consortia? *GIS World* 8 (8), 82-84 p.

Schmidt, T.W., Ault, J.S. and Bohnsack, J.A. (1999): *Fisheries and Essential Habitats,* http://fpac.fsu.edu/tortugas/studyarea/ault.html.

Vckovski, A. (1998) *Interoperable and Distributed Processing in GIS*, Taylor and Francis, London, ISBN 0748407928.

Wright, D.J., Fox, C.G. and Bobbitt, A.M. (1997) A Scientific Information Model for Deep Sea Mapping and Sampling, *Marine Geodesy* 20, 367-379 p.

Wright, D.J. and Goodchild, M.F. (1997) Data from the Deep: Implications for the GIS Community, *International Journal of Geographic Information Science,* 11, 5 p.

8

A Knowledge-Based GIS For Concurrent Navigation Monitoring

Frederic Barbe, François Gélébart, Thomas Devogele and Christophe Claramunt

The development of prescriptive and monitoring systems for maritime navigation can be suitably explored within a GIS environment. It provides important database, visual and analysis functions that can be usefully integrated within a computing environment that simulates and controls ship displacements. This paper introduces an experimental prototype that simulates the concurrent displacement of several ships, monitors navigation risks and triggers appropriate routing decisions. The simulation model integrates maritime bathymetric data within a GIS computing environment that acts as the kernel of the prototype developed.

The simulation is based on the notion of a user ship, controlled by the user, and external ships whose courses are randomly defined. Ships are mainly characterised by their positions, which are a function of time, and their intended course. Potential navigation conflicts include collision risks and running aground. The management of collision risks is based on the concept of closest point of approach defined from relative ship routes. The monitoring of these risks is based on a minimal distance approach in both the spatial and temporal dimensions. Collision avoidance is modelled from international navigation rules, constituting the knowledge-based component of the prototype. Such a computing environment provides a useful system for training maritime officers and for understanding navigation rules.

1   Introduction

Maritime GIS represents an emerging area of study whose applications cover the development of an integrated ocean management (NRC, 1997), study of submarine environment (Lehman, 1998) and integrated management of fisheries and marine activities (Lee and Kemp, 1998) to mention some recent examples. Another direction of important interest is the application of GIS for maritime navigation. Recent progress in this area includes the development of navigation-aid systems that combine maritime geographical data with Earth positioning systems (GPS), radar sensors (Rolfe and Alexander, 1997), and the scheduling of ship routes (SIAMS, 1999). The next generation of maritime GIS will evolve towards monitoring and decision-support systems that integrate temporal and geographical information within a collaborative telecommunication and computing environment. As such, these systems may act as proactive resources that facilitate concurrent navigation monitoring in overloaded maritime areas. The research described in this extended abstract introduces a GIS prototype whose objective is the simulation and management of collision risks within a maritime navigation environment. Our research aims are oriented towards a predictive decision-support system that improves real-time monitoring of maritime navigation.

The database support of this framework integrates bathymetric data and navigation information modelled at the level of individual ships. The dynamic component of the maritime database is based on concurrent simulation of several ship displacements and prescriptive monitoring of collision risks that prevent ship collisions or grounding. We distinguish between user ships and external ships, whose courses are evaluated in an almost random basis while the user controls the trajectory characteristics of his ships and adapts their courses during the simulation process. The monitoring component of our prototype is generated by a navigation simulation that evaluates next ship locations within a maritime region of study. This simulation estimates successive ship locations using navigation parameters. The dynamic knowledge of the simulated system is composed of individual and relative navigation rules. The former rules are given by the initial parameters of the navigation, the latter by international standards (IMO, 1997). The prototype is relatively open as the different initial parameters of the simulation can be predefined depending on user requirements (e.g., ship speed, course and location). The development of this real-time maritime prototype integrates the following software development components: (1) design and implementation of a temporal and geographical database, (2) representation and monitoring of the maritime system behaviour, that is, individual and relative ship displacements and risk assessments, and (3) design of a user-oriented interface. The remainder of this paper is organised as follows. Section 2 introduces the maritime database and the model used for its representation. Section 3 develops and motivates the method identified for the calculation of the closest point of approach. Section 4 presents the monitoring of collision risks. Section 5 introduces and illustrates the simulation processing. Finally section 6 draws the conclusions.

2   Maritime Database Representation

The Maritime data support of this research includes a bathymetric database provided by the French naval hydrographic and oceanographic service (SHOM). This bathymetric reference is based on coastal lines, bathymetric lines and soundings referenced in the three geographical dimensions. In order to provide a computationally efficient structure for the monitoring and simulation of ship displacements, that is, the depth in any estimated location, Voronoï polygons have been generated from the soundings geo-referenced as points. These polygons provide a partition of the maritime space that gives an immediate approximated depth in any maritime location, an important requirement of our real-time computing application. Figure 1 illustrates this partition applied to the region of study, the bay of Brest.

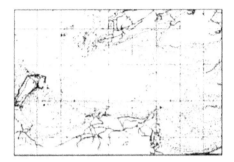 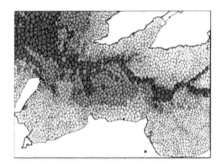

Figure 1. Punctual data vs. Voronoï bathymetric

The information represented within the geographical system integrates real and simulated data. The real component of the database is composed of the bathymetric database generated, and the description of the static and dynamic properties of the ships considered for the processing of the simulation. Our model represents user ships (one-to-many), and the external ships part of the maritime environment. The static properties of a ship consist of thematic attributes that characterise its properties (e.g., type, draft). The temporal properties of a ship include successive locations, each represented by a point and time-stamped by a temporal instant, displacement parameters (i.e., course and speed) between successive locations (considered as constant between two locations as a first approximation). The initial parameters of the simulation are defined as follows:

- Definition of the number of user and external ships;
- Initialisation of user ship properties (e.g., initial position(s) and intended route(s));
- Initialisation of external ship properties (as for user ship properties or randomly).

The simulation is generated from the concept of intended movement that is determined on a one-minute basis (i.e., discrete approach in space and time). The maritime database includes past, current, and estimated ship positions. The conceptual database is based on a modelling approach that associates the description of temporal entities to their spatio-temporal processes (Claramunt and Thériault, 1995; Claramunt et al. 1997); it is based on the MADS object-relationship model that integrates spatial and temporal concepts (Parent et al. 1998). Figure 2 introduces the database schema of the maritime application. This schema illustrates several modelling concepts (e.g., generalisation and aggregation abstractions, spatio-temporal processes represented as relationships, spatial and temporal entities). For example, a set of ships is defined as the aggregation of sailing ships, ferry-boats, warships, fishing ships and tankers. A generalisation relationship is defined between any ship (e.g., a ferry-boat) and a ship. Running aground and collision are defined as spatio-temporal relationships. An obstacle that leads to a running aground is due to either a coast line, an insufficient depth or a floating obstacle.

The conceptual schema, Figure 2, presents a visual view of the database that facilitates communication between application specialists and database designers. It also supports the mapping of the application towards a logical implementation within the GIS. At this stage, the different modelling concepts are translated to corresponding database constructors. The database implementation is based on a tuple-versioning approach that time-stamps thematic and spatial attribute values by a period of time (Lum et al., 1984). Figure 3 illustrates these principles by presenting a temporal table that describes the successive positions and time instants of three ships, their courses and speeds. Additional tables are generated for the management and representation of initial data and ship intended routes.

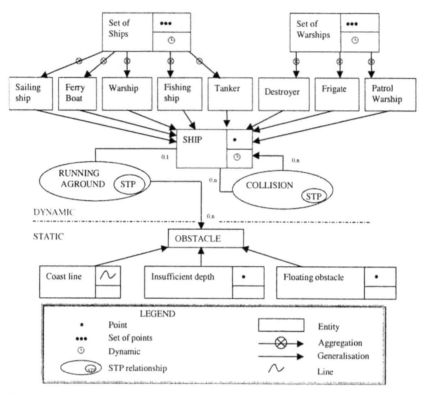

Figure 2. *Database schema*

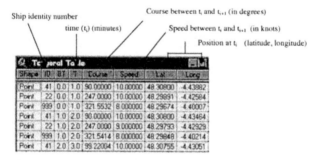

Figure 3. *Dynamic database representation*

3 Evaluation Of The Closest Point Of Approach

This section summarises the principles used for the determination of the Closest Point of Approach *CPA* between two ships. The *CPA* is defined as the minimal distance that separates two ships whose routes are crossing (Figure 5). We briefly introduce some maritime chart principles which are used for the computation of the *CPA*. A ship position is determined by its longitude G and latitude φ in degrees. Within a region of interest whose latitude extension is given by φmin and φmax, the distance D between two points

A and *B* is given by

$$D = \sqrt{(\varphi A - \varphi B)^2 + (GA - GB)^2 \cos^2(\varphi m)} \qquad (1)$$

where $\varphi m = \dfrac{\varphi min + \varphi max}{2}$

The azimuth Z between two points is the angular difference between the line defined from these two points and the geographical North whose direction is given by the meridians. As illustrated in Figure 4 left, the application of a trigonometric relation gives

$$Z' = \text{Arctan}\left(\frac{(GA - GB)\cos(\varphi m)}{(\varphi B - \varphi A)}\right) \qquad (2)$$

The azimuth Z is then derived thanks to the sign of $(\varphi B - \varphi A)$ and $(GB\text{-}GA)$ as illustrated in Figure 4 right.

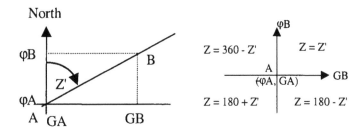

Figure 4. *Derivation of the azimuth*

The anti-collision evaluation is based on the *Closest Point of Approach CPA* as illustrated in Figure 5.

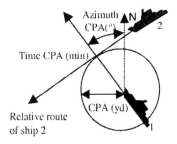

Figure 5. *Closest Point of Approach*

The *CPA* is estimated from the ship relative routes. In order to derive the *CPA*, the displacement vector of each ship is decomposed into two vectors in the latitude φ and longitude *G* axis as illustrated in Figure 6.

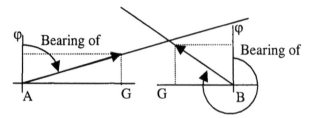

Figure 6. *Ship displacement vectors*

The functions that define the ship positions in function of time are defined as follows:

$$(t) = (SA \times t \times \cos(HA)) + \varphi Ai$$

$$(t) = \frac{SA \times t \times \sin (HA)}{\cos (\varphi m)} + GAi$$

$$(t) = (SB \times t \times \cos(HB)) + \varphi Bi$$

$$'(t)= \frac{SB \times t \times \sin (HB)}{\cos (\varphi m)} + GBi$$

Where $(\varphi Ai, GAi)$ and $(\varphi Bi, GBi)$ represent the initial positions of A and B, respectively; *HA* and *HB* ship headings; *SA* and *SB* ship speeds. By extension of the formula (1) to the temporal dimension, the function of time that defines the relative distance between the two ships is defined as follows:

$$D(t) = \sqrt{(\varphi A(t) - \varphi B(t))^2 + (GA(t) - GB(t))^2 \cos^2 (\varphi m)} \qquad (4)$$

The minimum distance is then found by derivation with

$$\varphi A'(t) = SA \times \cos(HA) = A1 \qquad \varphi B'(t) = SB \times \cos(HB) = B1$$

$$GA'(t) = \frac{SA \times \sin (HA)}{\cos (\varphi m)} = A2 \qquad GB'(t) = \frac{SB \times \sin (HB)}{\cos (\varphi m)} = B2 \qquad (5)$$

and $D'(t) = 0$ (CPA is the minima of the D(t) function), then

$$CPA = \frac{(A1 + B1)(\varphi Ai - \varphi Bi) + (A2 + B2)(GAi - GBi)(\cos^2 (\varphi m))}{(A1 - B1)^2 + (A2 - B2)^2 (\cos^2 (\varphi m))} \qquad (6)$$

From (3) and (6) and the definition of the azimuth in (2), the longitude and latitude, the distance, the temporal distance (denoted *TCPA*), and the azimuth of the *CPA* (denoted *ZCPA*) are obtained. Overall the *CPA* gives a fundamental parameter for the simulation of ship displacements and the evaluation of collision risks. Together with the analysis of running aground hazards, it constitutes a support for the management of navigation risks.

4   Risk Management Principles

Starting from the initial positions, the simulation evaluates several risks. A first control is the management of geographical risks (anti-grounding control): each ship location, either current or estimated, is analysed in function of the depth and ship draft (an unsafe location is either a location with an insufficient depth or a position on the ground). An additional constraint is applied on ship routes in order to avoid running aground between two successive ship locations. A second control concerns the evaluation of risks for concurrent ship displacements, so-called, anti-collision evaluation (Froese and Mathes, 1996). Figure 7 illustrates the functional approach taken for user ships. The simulator directly evaluates new speeds and headings that lead to a safe position if required. However, the user controls the validation, or not, of these changes as illustrated in Figure 7. On the other hand, external ships are directly guided according to international navigation rules.

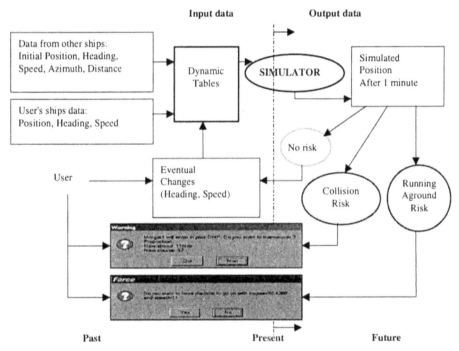

Figure 7. *Monitoring of user's ships displacements*

The monitoring of collision risks is based on a minimal distance of approach that is an application dependent parameter, fixed at 500 yards for demonstration purposes within the current version of our prototype. The simulation evaluates collision risks on a step by step basis and in ascendant *CPA* order, i.e., the higher collision risk. Collision avoidance is then triggered according to international navigation rules (IMO, 1997):

- When two ships are facing each other, they both have to go starboard (Figure 8 left);
- If they are not, the ship coming from starboard is privileged, so the ship coming from port has to change its course (Figure 8 centre);
- If a ship is overtaking a second ship, it is not privileged and has to change its course (Figure 8 right).

In order to modify navigation routes in the case of a collision risk between two ships, prescriptive rules integrate the time and azimuth of the *CPA*, and the relative distance and azimuth of the ships. The prototype identifies different conflict situations by an

Figure 8. *Anti-collision rules*

integration of the distance D and azimuth of the potentially dangerous ship, the *CPA* temporal distance (function of the relative ship speeds), denoted *TCPA*, and the azimuth of the *CPA*, denoted *ZCPA*. To follow international navigation rules, the prototype discriminates five different situations, which are described in Figure 9.

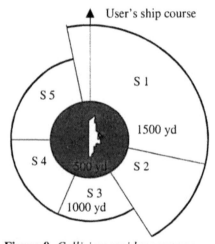

Figure 9. *Collision avoidance range*

The different situations presented by the above figure are monitored as follows:

- Case S1 – Bearing 350° to 100°: The external ship is privileged. The user's ship has to change its course if $D<$ 1500 or *TCPA* < 4 min. The new course is then *Course = Course* + 30°

- Case S2 – Bearing 100° to 165°: The external ship is privileged. The user's ship has to change its course if $D<$ 1500 or *TCPA* < 4 min. Two configurations are analysed: (1) If *Course* – 90° < *ZCPA* < *Course* + 90° then the new course is *Course = Course* - 45° (2) If *Course* + 90° < *ZCPA* < *Course* - 90° then the new course is *Course = Course* + 45°

- Case S3 – Bearing 165° to 195°: The user's ship is privileged. The user's ship changes its course if the external does not and if $D<$ 1000 or *TCPA* < 3 min. Then the new course is *Course = Course* + 30°

- Case S4 – Bearing 195° to 270°: The user's ship is privileged. It changes its route if the external ship   does not and if $D<$ 1000 or $TCPA <$ 3 min. Two configurations are analysed: (1) If $Course - 90° < ZCPA < Course + 90°$ then the new course is $Course = Course + 45°$ (2) If $Course + 90° < ZCPA < Course - 90°$ then the new course is $Course = Course - 45°$

- Case S5 – Bearing 270° to 350°: The user's ship is privileged. It changes its course only if the external ship does not and if $D<$ 1000 or $TCPA <$ 3 min. Then the new course is $Course = Course + 30°$

Figure 10 illustrates the case S1 in which the user's ship changes its course by 30 degrees. The above situations cover the extent of potential navigation conflicts between two ships. The complexity of the simulation augments with the overall number of conflicts. However, as most situations deal with a relatively limited number of conflicts, the proposed method can be considered as reasonably efficient.

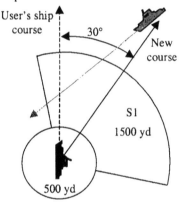

Figure 10. *Collision avoidance example*

5   Simulation Processing

The simulation processing is based on several initial parameters: number of user's ships and ship characteristics (i.e., location and intended movement), and external ship initial positions and routes (defined using either a random or selective mode by the user). The simulation process starts by verifying the coherence of each ship initial position. The depth relative to each ship location is compared to the ship draft. This control is then propagated to each calculated ship position during the simulation. When a ship depth is considered as insufficient, the simulator evaluates a new course in order to take in account the constraints of the maritime environment. A two-step interface approach is applied on user's ships and external ships. For user ships, a message is sent to the user whereas new courses are directly evaluated for external ships. These communication principles are illustrated by the Figure 11.

Once the next safe routes are identified, the simulation determines the successive positions of the ships and derived parameters (i.e., azimuth, heading, distance, speed). These parameters are stored within the simulation database. The temporal frequency of the current version of our prototype is currently fixed to one minute. A historical database keeps track of previous situations and serves as reference knowledge for future simulations (similar conflicts are resolved with similar navigation decisions). Such a historical database is particularly interesting when a ship is very close to the coast line: a previously calculated safe position can be directly applied to the ship route. The prototype computing environment has been developed within the ArcView GIS and the Avenue programming environment.

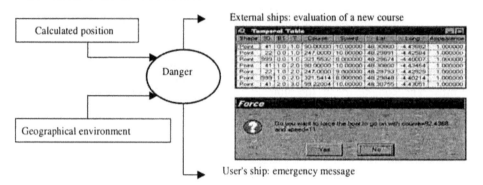

Figure 11. *Risk management interface*

The following figures are illustrative configurations that describe the management of either grounding or collision risks. Figure 12 left introduces an example of a route that might lead to a running aground. The simulation evaluates the displacement of the ship and determines successive safe positions. Figure 12 right presents an example of running aground monitoring between the routes of two ships.

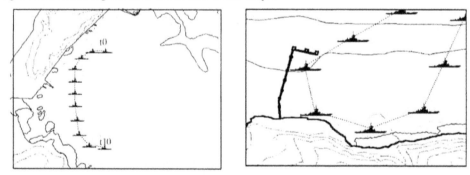

Figure 12. *Running aground monitoring examples*

Figure 13 presents another relevant example of anti-collision as the two ships simulated are close to the coastline. The ship at the bottom of the figure is the user's ship. In such a situation the two ships should manoeuvre to starboard. However, after a warning message, the system has forced the user's ship to keep its route. Therefore, the external ship manoeuvre, at the top of the figure, takes into account both the geographical environment and the user's ship route.

The principles of these simulations are based on several assumptions. First, the prototype supposes that all ships navigate according to international rules to avoid collision. Unfortunately, such an assumption does not always correspond to real

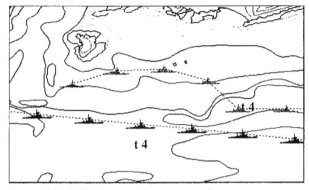

Figure 13. *Collision avoidance example*

situations. A second assumption of the system is the notion of closest danger in complex situations. The impact of a collision avoidance can be difficult to evaluate and simulate compared to other immediate danger situations. Such a complex situation often leads to several concurrent conflicts that are difficult to evaluate within a simulation environment. Indeed an experienced officer on watch could find an appropriate solution in such a situation. This range of situations is at the borderline between the limits of the potential of such a simulation system and human decision-making processes which are often difficult to simulate within a computing environment. Overall the GIS prototype, implemented on a laptop computer, simulates the system behaviour within an acceptable computing processing time (e.g., 1 minute for a 15- minute navigation).

6   Conclusion

The development of GIS applications is now facing new challenging domains. In particular, many application areas require the integration of the temporal dimension and the simulation of either natural or anthropic phenomena. Maritime GISs are one of these areas in which the development of real-time GIS might be of particular interest for navigation monitoring and the simulation of ship displacements. This paper introduces a novel approach for the simulation of ship displacements and the management of navigation risks. It is based on a maritime database that integrates bathymetric data and a simulation model that characterises ship properties and courses. The prototype developed integrates navigation rules as a knowledge-based component and triggers appropriate navigation decisions to avoid collision and running aground. The model has been applied on small navigation configurations that illustrate the potential of such a system for navigation monitoring. Such a prototype provides a suitable environment for the planning of navigation routes in overloaded maritime areas, and the training of maritime officers. Further work concerns the integration of (1) positional sensors such as radar and GPS systems and a collaborative integration with an Electronic Chart Display Information System (Rolfe and Alexander, 1997), and (2) universal shipborne Automatic Identification Systems (AIS) to adjust navigation decisions to ship characteristics (e.g., a tanker manoeuvre is slower than a fishing ship manoeuvre) (Alexander and Prime, 1999). The prototype might also be evaluated in overcrowded maritime regions, and as a support for monitoring commercial navigation.

References

Alexander, L. and Prime, K., 1999, Learning from Prince William Sound ; ECDIS and AIS as a natural synergy. In *Proceedings of the International Congress on Ship and Maritime Transport*, pp 169-176.

Claramunt, C. and Thériault, M., 1995. Managing time in GIS: An event-oriented approach. In Clifford, J. and Tuzhilin, A. (eds.), *Recent Advances in Temporal Databases*, Berlin, Springer-Verlag, pp. 23-42.

Claramunt, C., Parent, C. and Thériault, M., 1997, Design patterns for spatio-temporal processes. In *Searching for Semantics: Data Mining, Reverse Engineering*, S. Spaccapietra and F. Maryanski (eds.), Chapman & Hal, pp. 415-428.

Froese, J. and Mathes, S., 1996, Computer-assisted collision avoidance using ARPA and ECDIS. In *Maritime collision and prevention*, 2, pp 200-214.

Gilling, S. 1997, Collision avoidance, driving support and safety intervention systems. *Journal of Navigation*, 50(1), Bonnor (ed.), The Royal Institute of Navigation, Cambridge, UK, pp. 27-32.

IMO, 1996, *Code on Alarms and Indicators, International Maritime Organization Performance Standards for ECDIS (Appendix 5)*, technical report IMO-867E.

IMO, 1997, *International Safety Management Code (ISM Code) and guidelines on implementation*, technical report IMO-118E.

Judson, B., 1997, A tanker navigation safety system. *Journal of Navigation*, 50(1), Bonnor (ed.), The Royal Institute of Navigation, Cambridge, UK, pp. 97-108.

Lee, H. and Kenp, Z., 1998, Supporting complex spatiotemporal analysis in GIS. In *Innovations in GIS 5*, S. Carver (ed.), Taylor & Francis, pp. 151-161.

Lehman, A., 1998, GIS modelling of submerged macrophyte distribution using generalised additive models. *Plant Ecology*, 129, pp. 113-124.

Lum, V., Dadum, P., Erbe, R., Guenauer, J., Pistor, P., Walch, G., Werner, H., and Woodfill, J., 1984, Designing DBMS support for the temporal dimension. In *Proceedings of the SIGMOD'84 Conference*, (New York: ACM), pp. 115-126.

National Research Council (NRC), 1997, *Striking a balance: strengthening marine area governance and management*, Washington DC, National Academy Press.

Parent, C., Spaccapietra, S., Zimányi, E., Donini, P., Plazanet, C. and Vangenot, C., 1998, Modelling spatial data in the MADS conceptual model. In *Proceedings of the 8th International Symposium on Spatial Data Handling*, SDH'98, pp. 138-150.

Rolfe, G.A. and Alexander, L., 1997, Integrating ECDIS and radar: an overview. *Sea Technology*, 38(3), pp. 10-15.

SIAMS, 1999, *Ship Information and Management System*, Telematics Applications Programme, EU DGXIII C/E, http://www.siams.net/.

Smierzchalski, R., 1999, Evolutionary trajectory planning of ships in navigation traffic areas. *Journal of Marine Science*, 4(1), pp. 1-7.

9

Adapting One's Mental Model: An Essential Process For Successful Navigation In An Environment

Hartwig Hochmair

Navigating through an unknown environment is a common task that is often successfully completed even using an incorrect map. This paper proposes formalized strategies that describe how humans navigate through an unknown environment with the help of a partially incorrect map. The formalized strategies are based on the structure of an agent with state and presume that the human's mental representation of the environment adapts to the environment through perception of information from the real world. During the navigation process, initial information and perceived information are fused. Recent research is aimed at describing errors in mental representations. This paper takes a different approach and classifies several cases that result from differences between the real world and a map. It shows the agent's reactions and how its decision behaviour is influenced by the detected differences.

1. Introduction

A common task of humans is to navigate from one place to another, often in an unknown area. Typically, street network maps may be used. Most of the time successful navigation is possible, although the maps are partially incorrect. The term incorrect means that the topographical content in the map differs from the real world. The novel contribution of this paper is a proposed classification of decision situations, which occur during the navigation process and result from errors in maps. It describes what makes a situation decidable or not. The paper presents formalized navigation strategies for successful navigation cases in an unknown street network, even with incomplete maps.

1.1 Hypothesis

A situation is called critical if a decision about which road to take cannot be made with certainty. Critical situations result from differences between the real world and the agent's mental representation of the environment. The decisions of the navigating agent are based on a paper map that is partially incorrect. An incorrect paper map either omits an element of the real world (omission) or represents an element that does not exist in reality (commission). The navigating agent has to detect the incorrectness of the map and decide which of these two cases is the explanation for an error in the map. The hypothesis of the paper is that successful non-backtracking navigation is possible if the agent is able to find out whether omission or commission cause differences between the real world and its representation on a map. More complex analysis is necessary for cases where the destination is reached using backtracking (not covered here).

1.2   Background

Geographers have studied the cognitive process of map readers while doing research on symbolization and design principles for cartography (Robinson and Petchenik 1976). Experimental studies were conducted to explore the interaction between the map and its reader. Sheppard and Adams (1971) found that up to 50 percent of persons use road maps for route finding in an unknown area. Gray and Russell (1962) found that 49 percent of the car drivers keep a map in their car, 16 percent of them use the maps 'often' and 21 percent 'sometimes'.

The map an agent has is important to find routes in an environment, determine relative positions and describe the current location (Kuipers 1978). Casakin, Barkowsky et al. (forthcoming) point out that the agent requires a suitable representation of the environment to determine its position. Mental representations and cognitive maps are often considered to be split into several parts (Lynch 1960, Appleyard 1970, Beck and Wood 1976, Kuipers 1982). Acquisition of spatial knowledge, through exploration of the environment and using a combination of various information modes, was discussed (Siegel and White 1975, Montello 1993, Hutchins 1995). In a similar way to these ideas, the approach of this paper proposes a merging of different information sources to construct the mental representation: the (incorrect) information of the paper map and the perceived information during the navigation process.

Much research aims at finding errors in mental representations of the environment. Systematic differences between people's knowledge of the world and the reality, such as hierarchical spatial reasoning (Stevens and Coupe 1978), perceptual organizing principles (Tversky 1981), or a varying perspective (Holoyak and Mah 1982) are found. Lynch (1960) points out that errors in cognitive maps are most frequently metrical, and rarely topological. We propose that topographical errors in combination with metrical errors can lead to critical navigation situations. In the literature, the consequences for the navigation process when maps are partially incorrect, are not discussed.

1.3   Agent and Environment

There are many definitions of an agent. Many of these definitions point out that an agent is something that can perceive its environment through sensors and act upon the environment through effectors. Several abstract architectures for intelligent agents are introduced by Wooldridge (1999) and Genesareth and Nilsson (1987). From the different architectures, the type 'goal based agent with state' - contrary to 'purely reactive agent' - seems to be useful to describe a person's navigating and decision process in an environment. For our approach we use simulated agents throughout. An agent with state has some internal data structure, which is typically used to record information about the environment state and the history of actions. The components of the agent's state can be divided into those which are invariant (static) and those which are changeable (dynamic) during the navigation process.

Table 1 describes the elements of an agent's state. Barfield (1993) proposes three types of transitions in a system from one state into another: Time transitions, system transitions and user transitions. We assume that the agent's state is exclusively changed by the agent's actions, such as moving and perceiving, and is invariant to time or other external effects, therefore corresponding to the concept of user transition.

Russell and Norvig (1995) offer a list of properties of environments, e.g., accessible vs. inaccessible or deterministic vs. nondeterministic and give examples for environments with different properties. The authors call an environment dynamic if it can change when the agent performs a deliberate action on it. The environment is static if it cannot change through actions of the agent. We map the properties dynamic and static to the representation of an environment: If the agent's mental representation of an environment changes through movement or perception of the agent, it is called a dynamic mental model. In contrast to this, an environmental representation which is not influenced by the agent's behaviour, is a static model, e.g., the information on a paper map of a street network.

1.4   Mental Representations

For navigating an environment with the help of a paper map we need to specify which type of mental representation is useful for this task. Contrary to split representations, such as cognitive collages (Tversky 1993), we assume that the mental representation of the street network is one single part, as the required information is taken from one single paper map. In addition to spatial mental models, which include categorical spatial relations among elements, cognitive maps (Tolman 1948) unify landmarks and routes with metric survey information. Metric information is an important part for making decisions in the proposed mental behaviour model. Hence for our task a cognitive map seems to fit best for the description of the mental representation of the environment. The dynamic cognitive map represents the complete knowledge of the agent about the environment, may it be spatial or non-spatial. Technically it can be called 'belief' to stress the potential for differences between reality and the agent's believes about reality (Davis 1990).

2. Navigation In A Real Environment

At the beginning of the navigation process, the virtual agent gets the information from reading a partially incorrect street network map, which results in a partially incorrect cognitive map. As all of the agent's navigation decisions are based on the dynamic cognitive map, decision errors during the navigation process can occur. From the dynamic cognitive map, the agent extracts a planned path from the start point to the end point. It is a sequence of street segments and crossings, somewhat similar to Kuipers' TOUR model (Kuipers 1978). Through observations during the navigation process the agent can detect differences between the environment and the real world. The agent has to deal with these differences, which can result in an adaption of its believed position, the dynamic cognitive map, or the planned action and the planned path.

According to an inclusion of static and dynamic components in a representation of an environment, we use the structure of Table 1 for the real environment and its representation in the agent's state. The property 'static' means invariance of elements during the whole navigation process.

Real environment	static	dynamic
Topography of the street network	x	
Agent's position in the real environment		x
Agent's state		
Cognitive map adapted during the navigation process		x
Planned path from the actual position to the endnode		x
Agent's believed position		x
Planned action as part of the planned path		x

Table 1.   *Static and dynamic components in the real environment and the agent's state*

1.5   Navigation Process

The mental procedure of a navigating agent is embedded in a repeated process, until the agent has reached the target. The navigation process consists of following steps:

1. The agent has the **internal state** i_n which contains the dynamic cognitive map, the agent's believed position in the environment, the planned path, and the planned action.

2. Reaching an intersection, the agent **observes** the environment (indicated through the *see*-function in Fig. 1) which is restricted to the agent's actual field of view.

3. The observed geometry and the geometry stored in the cognitive map are **compared**, which can lead to differences (omission, commission).

4. The agent's state and observations form a *next*-function, which maps these two sources to an **updated state** i_{n+1}. The update can result in a change of the agent's cognitive map, the believed position, the planned path and the planned action.

5. The agent **performs** the (changed) planned action and updates the believed position in that state. The navigation process is terminated if the assumed actual position is the target node.

1.6   Detecting Differences which Result in Problems

Differences between the paper map and the real world are caused either by omission or commission (we exclude metric errors that result from geometric inaccuracy of the map in this approach). Omission describes the case that the agent finds unexpected elements in the real world, which are not drawn in the paper map, commission is the case if elements are drawn in the map but do not exist in the real world. Problems in the decision process can have three reasons and occur if the agent
(a) does not realize omission and commission at all,
(b) cannot find out whether omission or commission cause the detected differences,
(c) does not recognize any known topographic element in the cognitive map.

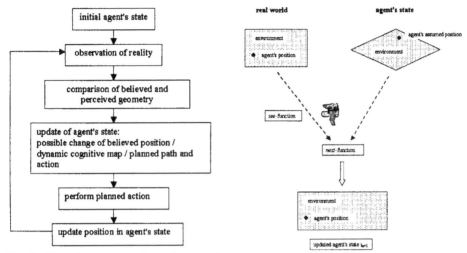

Fig. 1. *Mental steps to update the agent's state*

The first two cases result in the agent's inability to determine its position. We call such a case which results in decision problems a critical decision case.

Case (a) can occur if the geometry of a new detected intersection is similar to the expected one and the agent is not able to determine from the covered travel distance, whether the found intersection is that expected or a new one. The second case happens if due to an omission a new unexpected intersection is encountered that has a similar geometry to an expected intersection. The agent then realizes the difference but cannot determine its position. The third case occurs if an unexpected element is detected that shares almost the same direction as the actual street segment of the planned path.

Detecting a difference, the agent makes a hypothesis about which effect has caused the differences. A correct hypothesis is of high importance for successful navigation, as it influences the agent's ability to determine its actual position. In case (a) the agent does not make a hypothesis, in cases (b) and (c) the agent is unable to make the right hypothesis.

In addition to the descriptions of potential problems we give an example as demonstration (case b): The agent's planned path leads from node N 10 via node N 1 to node N 2. Fig. 2a shows the agent's incorrect cognitive map taken from an incorrect paper map. Finding an unexpected street segment on the edge e10,1 (Fig. 2b) can lead to the same geometrical situation as a missing street element at node N 1 (Fig. c). If the agent cannot estimate the covered distance from node N 10 with a sufficient accuracy, it cannot determine the actual position and is unable to make the correct hypothesis. Commission can also be caused by a perceptive error, such as not recognizing a street segment at an intersection.

We claim that in most decision cases the agent can realize whether omission or commission is the reason for differences between the real world and its mental representation, although the agent's assumption cannot be verified until it reaches the next known intersection or landmark.

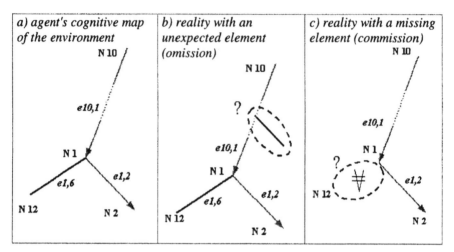

Fig. 2 *a,b and c: Omission and commission causing a critical navigation situation (case b)*

1.7 Classification

We create a classification of decision situations, which can happen during the agent's navigation process. Corresponding to each situation, the agent performs a specific mental activity to successfully continue navigation. The situations are classified based on the detected differences and depend on the following three parameters: the number of available street segments that are part of the planned path, the number of available street segments that lead to a node of the planned path, and the criterion if the actual node was part of the cognitive map before the actual move. Fig. 3 shows which criteria at the decision points in the mental behaviour model lead towards which decision case.

Table 2 shows the specific decision situations and the type of error, which causes them. The potential errors are omission, commission or perceptive errors.

We decide between successful and unsuccessful cases. We call a case successful if the agent is able to continue the navigation process, can determine its position and realize known elements at an intersection. Assumptions about omission and commission can definitely be confirmed after reaching a known and expected intersection. Critical case (0.0) happens if one of the three situations mentioned in section 1.6 occurs. In this case a correct decision to successfully continue the navigation process is not possible.

In cases 1.1.1 and 1.1.2 the perceived street segments do not contain an element of the planned path, but can lead to a node of the planned path.

Case 1.1.1 (shortcut)
The dynamic cognitive map is completed with an omitted street segment, which leads to a node of the planned path. The agent finds a new planned path, which is shorter than the old one.

Case 1.1.2 (detour)
Due to a committed street segment the agent is forced to determine a new planned path using a detour. A detour can only then be found if the committed street segment represents an edge in a root graph. The new planned path leads through a node that was not included in the former planned path.

Fig. 3. *Agent's mental behaviour after reaching an intersection*

Description	caused by	Case
Critical decision case		
Agent cannot identify/distinguish between omission and commission. Agent cannot identify known elements.	Omission/commission	(0.0)
Successful decision case		
Shortcut	Omission	(1.1.1)
Detour	commission	(1.1.2)
Unexpected intersection on street segment (splitting)	Omission	(1.2.1)
Agent does not see expected crossing; this can be formally equal to normal agent's move.	perceptive error results in commission	(1.2.2)
Planned path can be continued.	evt. omission/commission	(2.0)

Table 2. *Critical and successful decision cases*

In cases 1.2.1 and 1.2.2 one of the perceived street segments is an element of the actual planned path.

Case 1.2.1 (splitting)
The actual street segment is split into two street sections through omission of an intersecting street segment. The planned path is adapted through splitting of the actual street segment, but keeps the same overall length. The agent's coordinative position is unchanged although a new node is added to the dynamic cognitive map. The agent's position gets a new name as it is situated on the new node.

Case 1.2.2 (skipping)
Due to a perceptive error the agent does not recognize an intersection and therefore skips a road segment of the planned path undeliberately. The agent is able to determine its position through the network geometry or a landmark.

In case 2.0 (**continue planned path**), changes in the dynamic cognitive map are potentially caused through the perceived information. These changes do not influence the agent's believed position, planned path, or planned action. Therefore the planned path can be continued.

1.8   Illustrating Example

The following example demonstrates the consequences of omission and commission during the navigation process. The real world and the agent's dynamic cognitive map are schematized through nodes and edges. As mentioned by Mark, Freksa et al. (1999), the schematization of graphics often parallels the schematization of the mind, therefore a representation of a map and the real world through a graph seems useful.

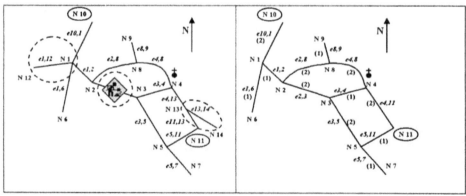

Fig. 4. *Street network in the real world*   **Fig. 5.** *Street network represented on a paper map*

Fig. 4 and Fig. 5 show a street network in the real world, and its representation in a partially incorrect paper map. Differences between these two maps are marked through circles. The closed road results in a commission error; the other two street segments are a consequence of omission errors. The nodes that result from omission and are therefore detected by the agent during the navigation process are labelled N 12, N 13 and N 14. The lengths of the street segments on the paper map are added in brackets.

The following figures visualize the agent's states at the nodes along the agent's planned path. On nodes N 1, N 2 and edge e4, 11 adaptations of the agent's state have to be done as a consequence of perceiving information from the environment during the navigation process.

Fig. 6 shows the agent's initial state at the start node. The planned path to the target node N 11 is indicated as a directed graph with arrows, whereas the rest of the cognitive map is visualized as an undirected graph.

Fig. 7 shows the agent's state after moving to node N 1. The agent recognizes a new street segment from N 1 to N 12 as a result of omission and adds it to a dynamic cognitive map. This new information does not influence the agent's planned path. The agent is able to determine its position, the planned path stays unchanged, there is no unexpected intersection, and no intersection was missed through a perception error during the last move. Hence the expected planned path can be continued (case 2.0).

At node N 2 (Fig. 8) the agent recognizes a closed street segment between N 2 and N 3 as a commission error. The closed street segment is part of the actual planned path; hence a detour has to be taken (case 1.1.2). The observations in node N 2 result, besides a change of the dynamic cognitive map, also in a change of the planned path.

Fig. 9 shows a situation representing case 1.2.1 (splitting). Along the street segment from N 4 to N 11 the agent detects the new intersecting road e13, 14. The new node N 13 is added in the agent's cognitive map, the actual position is updated to N 13. In its cognitive map, the agent splits the actual road segment e4, 11 into the edges e4, 13 and e11, 13 and adds edge e13, 14 starting from the new intersection. As the agent recognizes edge e13, 14 as a result of omission, no critical situation occurs. The agent recalculates the planned path, which then consists of the edge e11, 13. After one more move the agent reaches the target node N 11.

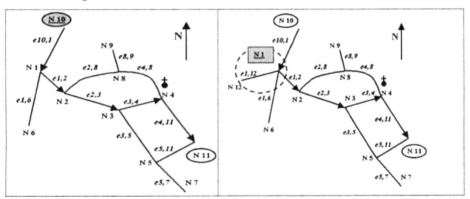

Fig. 6. *Agent's dynamic cognitive map and planned path at starting node N 10*

Fig. 7. *Agent's dynamic mental model and planned path at node N 1*

3. Formalization

This section describes a simulated agent's behaviour during the navigation process as well as the environment that is navigated. The method used for formalization are algebraic specifications, which are implemented through the functional programming language Haskell (Thompson 1996). The purpose of a specification is to formally describe the

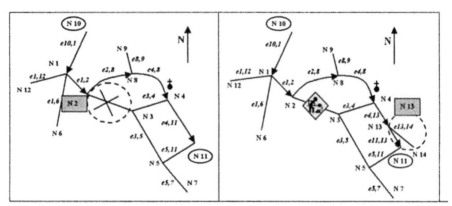

Fig. 8. *Agent's cognitive map and planned path in node N 2*

Fig. 9. *Agent's cognitive map and planned path in new node N 13*

behaviour of objects. The algebraic specifications offered here include formalized strategies about how the agent deals with errors in maps, and describe the ontology of the agent's cognitive map. Algebraic specifications are used to formally prove the correctness of the implementation of the presented model. Consequences of errors in the map, or errors in the agent's perception of reality can be modelled, and the behaviour of the agent can be predicted.

The model contains

- A static environment, which stays invariant during the agent's navigation process
- A map reading and navigating agent which navigates through the environment and adapts its representation of the world through the perceived information during the navigation process

The environment is abstracted to a set of connected nodes and edges, where each edge has a certain length. It is built so as to include only the important elements for the way finding process. Hence one-way restrictions and turn-restrictions are excluded. All street segments of the real world are assumed to be labelled with their start and end node.

The agent is always aware of which nodes are connected to the actual position node. Whereas the static environment is invariant during the navigation process, the agent's state is updated after each move. The agent's activities are reduced to perceiving information from and moving through the environment. The agent's initial information is provided by a paper map, which contains a geometrical structure of the environment and its landmarks. The task of the map-reading agent is to navigate between two given nodes. The agent finds the shortest path to the destination using its cognitive map.

In the following sections abstractions of agent and environment are given. They consist of algebras (as *classes*) with the operations using the parameterized data types. In the class declaration, the first line (called the *class header*) states that class is defined and lists the parameters. In the following lines the signatures of operations are given, describing the types of their arguments and results. Classes that are not restricted to a specific data type contain *polymorphic* functions. These functions can be applied to different data types.

Instances connect the data types with classes as they define how the operations defined in a class are carried out using this particular representation.

1.9 Environment

A street network can be represented as nodes and connecting edges. The nodes represent intersections; the edges represent the street segments. As the class describing the environment includes nodes and edges it is given the name 'Graphs' in the class declaration.

For the construction of the environment, four user defined data types are declared. They are defined by the keyword *data* together with type constructors. A type constructor is a function that constructs a new data type from other predefined data types, such as Int, Float or String. We need to define a new data type for the cost of an edge, a node, an edge with startnode, endnode and the cost, and a graph that consists of the next highest node in the graph and a list of edges.

```
data Cost = Cost Float
data Node = N Int
data E n = E n n Cost    -- start node, end node, cost
data G e n = G n [e n]   -- next unused node in graph, list of edges
```

The algebra for the environment must contain functions that insert an edge into an environment, delete an edge from an environment, get a list of all nodes in the environment, and split the edge between two nodes. The shortest path algorithm requires a function that gets the length of an edge between two given nodes.

```
class Graphs g e n where
 insertG :: e n -> g e n -> g e n -- insert edge in graph
 deleteG :: e n -> g e n -> g e n -- delete edge in graph
 nodes :: g e n -> [n]            -- get list of nodes in graph
 cost :: n -> n -> g e n -> Cost -- get cost between nodes
 split :: (n,n) -> g e n -> g e n  -- splitting between two nodes
```

The class instance gives function definitions for the specific data type G E Node of a graph. The function deleteG uses a filter function that consists of a property and a list of elements. It returns those elements of the list, which fulfil the given property. In this instance the filter function returns those edges in a graph that are not equal to a given edge e. The function incNode (used in insertG) is member of the class Nodes and increments the integer value of a node name. It is implemented to increase the next free node of a graph when a new edge is inserted into the graph.

```
instance Graphs G E Node where
 insertG e (G n es) = G (incNode n) (e:es)
 deleteG e (G n es) = G n (filter (not . (equalEdge e)) es)
 nodes (G n es) = nub (concat [ nodesE e | e <- es])
 cost n m (G ma es) = cost' n m es
          where cost' n m (e:es) = if isAB n m e then costE n m e
                            else cost' n m es
          cost' n m [] = maxCost
```

The split function (used for case 1.2.1) has as input a pair of nodes (n1, n2), between which the splitting is to be done, and a graph. The result type is a graph. Four steps are included in the function:

- split the actual street segment into two segments (twosteps). This includes taking the cost of the actual street segment (oldCost), dividing it into two equal costs (halfCost) and assigning the new lengths to each of the new parts,
- add the new observed edge in the dynamic cognitive map, where its length is set 10,
- delete the old edge that is split.

```
split (n1, n2) (G next es) = deleteG (E n1 n2 0) g3 where
    g3 = insertG (E next newNode1 (Cost 10.0)) g2
    g2 = insertG (E next n2 half Cost) g1
    g1 = insertG (E n1 next half Cost) (G next es)
    old Cost = cost n1 n2 (G next es)
    half Cost = divCost 2.0 old Cost
    newNode1 = incNode next
```

To create an environment we apply the insertG function to an edge of the environment and a graph that consists of a next free node N 0 and an empty planned path. Through the use of insertG the next free node of the graph is raised. We use the folder function to apply insertG to a list of edges of an environment. We define the function makeE which creates edges.

1.10   Agent

The agent's state consists of the actual position, a target node, a cognitive map of the environment and a planned path. For the definition of the agent we use two kinds of new types

- type synonyms (Position and Goal in our model)
- user defined types (PlannedPath and CognitiveMap in our model)

Type synonyms are an alias for an already existing data type; whereas user defined types can consist of several predetermined types in one single object. An agent's possible data structure is given below .

```
type Position = Node
type PlannedPath = [Node]
type Goal = Node
type CognitiveMap= G E Node

data Agent = Agent Position Goal CognitiveMap PlannedPath
```

At its actual position the agent can detect new street segments or realize the omission of street segments. The function observeEdge adds an observed edge to the dynamic cognitive map, function missingEdge deletes an edge from the cognitive map, the move-function changes the agent's position to the next node in the planned path. If the agent detects an unexpected intersecting street segment on the actual edge, it performs case (1.2.1) which is formalized in the splitting- function. All the functions are followed by a recalculation of the planned path, using the function getPathFromTo. It finds the shortest path between two nodes, following the algorithm of Dijkstra (Dijkstra 1959). The class declaration of the class 'State' includes the undefined parameter 'agent'; the following lines show the signatures of the included operations with their arguments and results.

```
class State agent where
    observeEdge :: E Node -> agent -> agent
    missingEdge :: E Node -> agent -> agent
    move :: agent -> agent
    splitting :: agent -> agent
```

The class instance applies the four class functions for the defined type 'Agent'. The function splitting includes the function fstSndNode which extracts the first two elements from the planned path and puts them into a pair of nodes, which is part of the input for the split-function.

```
instance State Agent where
    observeEdge e (Agent pos goal cogmap plan) = Agent pos goal newcogmap newplan
    where
      newcogmap = insertG e cogmap
      newplan = getPathFromTo pos goal newcogmap

    missingEdge e (Agent pos goal cogmap plan) = Agent pos goal newcogmap newplan
    where
      newcogmap = deleteG e cogmap
      newplan = getPathFromTo pos goal newcogmap

    move (Agent pos goal cogmap plan) = Agent newpos goal cogmap newplan
    where
      newplan = drop 1 ( getPathFromTo pos goal cogmap )
      newpos =  head newplan

    splitting (Agent pos goal cogmap plan) = Agent newpos goal newcogmap newplan
    where
      newcogmap = split (fstSndNode (getPathFromTo pos goal cogmap)) cogmap
      newplan = drop 1 ( getPathFromTo pos goal newcogmap )
      newpos = head newplan
```

At the start node the initial agent has an empty list for planned path as the planned path is recalculated at each of the agent's activities.

4. Case-Testing

This section introduces a virtual agent that navigates through the environment explained in section 1.8 and whose behaviour follows the introduced algebraic structure.

We create an agent at start node N 10 with target node N 11. The agent's cognitive map at the start node is called cmap and imagined to be taken from a paper map. It corresponds to the undirected graph in Fig. 5.

```
peter :: Agent
peter = Agent position endnode cmap []
  where position = (N 10)
      endnode = (N 11)

cmap :: G E Node
cmap = foldr insertG (G (N 0) []) [(makeE 10 1 2), (makeE 1 6 1), (makeE 1 2 1),
  (makeE 2 3 2), (makeE 2 8 2),(makeE 8 9 1), (makeE  3 5 2), (makeE 3 4 1),
  (makeE  4 8 2),(makeE 5 11 1), (makeE 4 11 2), (makeE 5 7 1) ]
```

The virtual agent moves from the start node to the node N 1. This results in an update of the agent's state: The position is changed to node N 1, the goal is still node N 11, the recalculated next highest node in the environment is node N 12, no changes in the cognitive map are made (indicated by three dots in the output of the calculation). The planned path stays unchanged and consists of the nodes N 1, N 2, N 3, N 4 and N 11.

Test input> move peter

>> position N 1 goal N 11 [next N 12, ...] plannedPath N 1, N 2, N 3, N 4, N 11

At node N 1 the virtual agent observes the new edge e1,12. Through this observation the new edge e1,12 (called e112 in the program) is added to the dynamic cognitive map (see Fig. 10), the planned path and the target stay unchanged, the next highest node is changed to node N 13 as node N 12 is now part of the cognitive map.

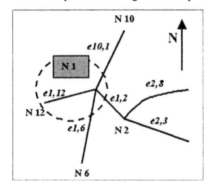

Fig. 10. *Change of the agent's cognitive map at node N 1 through observation of a new edge*

Test input> observeEdge e112 (move peter)

>> position N 1 goal N 11 [next N 13, from N 1 to N 12 cost 2.0,...] plannedPath N 1, N 2, N 3, N 4, N 11

At node N 2 the virtual agent realizes that edge e2, 3 is closed. The edge e2, 3 is skipped from the dynamic cognitive map (see Fig. 10), the recalculation of the planned path results in a change through a forced detour (see directed graph in Fig. 8). The target and the next highest node stay unchanged. To simulate the situation in node N 2 before realizing the missing edge, we create a virtual agent with a state at node N 2 and call him peter2.

Test input> missingedge e23 peter2

>> position N 2 goal N 11 [next N 13, ...] plannedPath N 2, N 8, N 4, N 11

During the next two moves no additional edges are observed. The virtual agent can continue its planned path to node N 4. In the middle of edge e4, 11 the agent observes a new intersecting street segment, which can be formalized through splitting (case 1.2.1). The observation causes a change of the agent's position to the new intersection N 13; the new edge is added to the cognitive map (see Fig. 12), the next highest node is node N15, the planned path changes through splitting of edge e4, 11 into the sequence N 13, N 14 (see Fig. 9). For the simulation agent 'peter' is moved to node N 4 and called peter4.

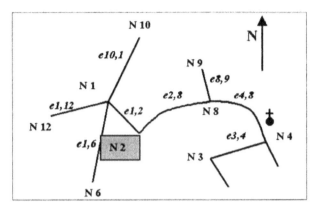

Fig. 11. *Change of the agent's cognitive map and planned path at node N 2 through realizing a missing edge*

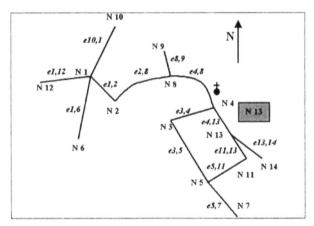

Fig. 12. *Change of the dynamic cognitive map and planned path at node N 13 through observation of a new intersecting street segment*

Test input> splitting peter4

>> position N 13 goal N 11 [next N 15, from N 13 to N 14 cost 10.0, from N 4 to N 13 cost 1.0, from N 13 to N 11 cost 1.0, ...] plannedPath N 13, N 11

From node N 13 the agent performs one more move to reach the target node N 11.

5. Conclusion And Future Work

This paper proposed a formalized decision behaviour of a navigating agent in an unknown environment using an incorrect paper map. It offered a classification of errors in maps and the resulting decision cases that can occur when navigating with the help of a partially incorrect map. It was shown that navigation decisions could only be made with certainty if the agent realizes omission and commission and is able to determine which of these two effects causes the detected differences. This criterion is equal to the agent's ability to determine its actual position. A critical case also occurs if the agent cannot recognize any known element at an intersection. The described and formalized decision cases are based on the idea that information from two sources is merged during the navigation process: the information from the agent's cognitive map and the perceived information.

As the formalized strategies are widely independent from the type of environment, future work will aim at finding similarities in wayfinding strategies and decision situations between a real street network and a non-Euclidean environment, such as the World Wide Web.

References

Appleyard, D. (1970). "Styles and methods of structuring a city." *Environment and Behaviour* **2**.

Barfield, L. (1993). *The User Interface - Concepts & Design.* Wokingham, England, Addison-Wesley Publishing Company.

Beck, R. and D. Wood (1976). Comparative developmental analysis of individual and aggregated cognitive maps of London. *Environmental Knowing: Theories, Research, and Methods.* G. T. Moore and R. G. Golledge. Stroudsburg, PA, Dowden, Hutchingson and Ross.

Casakin, H., T. Barkowsky, et al. (to appear). "Schematic Maps as Wayfinding Aids." .

Davis, E. (1990). *Representations of Commonsense Knowledge.* San Mateo, California, Morgan Kaufmann Publishers.

Dijkstra, E. W. (1959). "A note on two problems in connection with graphs." *Numerische Mathematik* **1**: 269-271.

Genesareth, M. R. and N. Nilsson (1987*). Logical Foundations of Artificial Intelligence.* San Mateo, CA, Morgan Kaufmann Publishers.

Gray, P. G. and R. Russell (1962). "Driver's understanding of road traffic signs." Social Survey Report.

Holoyak, K. J. and W. A. Mah (1982). "Cognitive reference points in judgements of symbolic magnitude."*Cognitive Psychology* **14**: 328-352.

Hutchins, E. (1995). *Cognition in the Wild.* Cambridge, MA, MIT Press.

Kuipers, B. (1978). "Modeling Spatial Knowledge." *Cognitive Science* **2**.

Kuipers, B. (1982). "The 'Map in the Head' Metaphor." *Environment and Behaviour* **14**: 202-220.

Lynch, K. (1960). *The image of the city.* Cambridge, Mass., MIT Press.

Mark, D., C. Freksa, et al. (1999). "Cognitive models of geographical space." *IJGIS* **13**(8): 747-774.

Montello, D. R. (1993). Scale and Multiple Psychologies of Space. *Spatial Information Theory: Theoretical Basis for GIS.* A. U. Frank and I. Campari, Springer-Verlag. **716**: 312-321.

Robinson, A. H. and B. Petchenik (1976). *The Nature of Maps.* Madison, WI, The University of Wisconsin Press.

Russell, S. J. and P. Norvig (1995). *Artificial Intelligence - A Modern Approach.* London, Prentice-Hall International, Inc.

Sheppard, D. and J. M. Adams (1971). "A survey of drivers' opinions on maps for route finding." *The Cartographic Journal* **8**: 105-114.

Siegel, A. W. and S. H. White (1975). The development of spatial representations of large-scale environments. *Advances in child development and behaviour*. H. W. Reese. **10:** 9-55.

Stevens, A. and P. Coupe (1978). "Distortion in judged spatial relations." *Cognitive Psychology* **10**: 422-437.

Thompson, S. (1996). *Haskell - The Craft of Functional Programming*. Harlow, England, Addison-Wesley.

Tolman, E. V. (1948). "Cognitive maps in rats and men." *Psychological Review* **55**: 189-208.

Tversky, B. (1981). "Distortions in memory for maps." *Cognitive psychology* **13**: 407-433.

Tversky, B. (1993). Cognitive maps, cognitive collages, and spatial mental model. *Spatial Informaition Theory: Theoretical Basis for GIS*. A. Frank and I. Campari. Berlin, Springer. **716**.

Wooldridge, M. (1999). Intelligent Agents. *Multiagent Systems - A modern Approach to Distributed Artificial Intelligence*. G. Weiss. Cambridge, Massachusetts, The MIT Press.

Section 3

Analysing the information: Computation and Modelling

10

Implementation Of A GIS-Based Cellular Automaton Approach For Dynamic Modelling

Verena Gruener

This chapter discusses an implementation of a Cellular Automaton approach to dynamically model plant dispersal. *Rhamnus Alaternus,* which is an invasive species, represents a threat to local ecosystems. The model can be used to achieve a better understanding of the processes involved in plant dispersal and aims at being a tool for environmental monitoring. An integrated approach combining Cellular Automata and GIS offers the ability of both spatially and temporally dynamic modelling as well as efficient data storage and analysis. The Cellular Automaton was simulated using the ARC/INFO Grid module and its programming language. It aims at delivering an exploratory tool to analyse the range of input parameters to environmental modelling and their effects on plant dispersal. The article discusses the properties of a CA and the general model structure. It further explains the main implementation steps, the problems encountered and solutions chosen. A discussion of the performance of ARC/INFO Grid in respect to the simulation of the CA concludes the paper.

1. Introduction

Truly dynamic modelling requires the consideration of both spatially and temporally dynamic processes. Contemporary GIS are considered to be poor performers in that respect, as they have a poor ability to handle dynamic spatial models and the temporal dimension (Wagner, 1996). Cellular Automata (CAs) enable explicit dynamic modelling, as they operate in discrete time steps, which model a spatial pattern that evolves over time. Due to its raster cells and rule-based neighbourhood relations a CA is of intrinsic geographical nature (Phipps & Langlois, 1997), It incorporates the ability to 'memorise' past states and easily handle the temporal variables, which enable the CA to model the temporal dimension (Wagner, 1997). A Cellular Automaton consists of a cellular lattice, where every cell has a potentially different state. The cells change their state over time according to the states of their neighbouring cells as well as so-called transition rules. However, the functionality for data input, storage and display is limited. Therefore, a combined Cellular Automaton and GIS approach promises to deliver a powerful modelling tool as well as efficient data handling. An integration of a CA within a GIS was considered most appropriate due to the wider availability of GIS. However, most work done on the integration of CA and GIS so far is of theoretical nature and only little literature on actual implementations can be found.

In the work presented in this paper, a Cellular Automaton was implemented within a GIS in order to model plant dispersal. The invasive plant species *Rhamnus Alaternus*, which is a native shrub of the Mediterranean, represents a threat to the native vegetation in New Zealand. Endangered native plants like the Pohutukawa tree now have to compete with the more resistant invasive species. The reproduction and spreading mechanisms of *Rhamnus Alaternus* are not fully understood yet. Concepts to limit the effect of *Rhamnus* on the local environment however need these ecological input parameters. Therefore, a better understanding of the processes involved in the dispersal of the plant is important. The combined Cellular Automaton/GIS approach promised to deliver a tool to provide better insight into these processes. Unlike complex plant dispersal models, a CA is able to simulate complex behaviour with only few input parameters. The developed model is applicable to other plants with similar behaviour as well and can therefore be used in environmental planning or ecological assessments.

ARC/INFO Grid has been chosen as the programming environment for the implementation of the Cellular Automaton. ARC/INFO is one of the most widespread GIS software packages. It offers a raster based tool with the Grid module as well as a programming language, which enables spatial data manipulation. Basis of the implementation is an existing conceptual plant dispersal model (Cole, 1999). An implementation of the Cellular Automaton was achieved and results show that the CA can be used as an exploratory tool where the effects of different parameter values on the plant dispersal can be tested.

An overview on Cellular Automata will be given after an introduction to the research problem. The general structure of the model is explained and the specifications for cells, states, neighbourhood and rules of the CA are shown. An explanation of the main features of the implementation and the problems encountered and solutions chosen follows. A discussion of the Cellular Automaton performance concludes this paper. The technical character of this paper aims at delivering a basis for the development of other CAs according to the concept presented in this paper.

2.   Cellular Automata

Formally, CAs are composed of the four elements *cell, state, neighbourhood* and *rule*. A CA consists of a regular uniform lattice, which may be of any extent. The square *cells* of the lattice make up the cellular space (Wolfram, 1986). Every cell has a discrete variable. The values of the variables of each cell specify the *state* of the cellular automaton. In a simple example, the cells of a CA have only two states, either the value 0 or 1. In general, the cells can take any finite number of possible states.

The state of a cell depends on the states of other cells in the neighbourhood. The *neighbourhood* is defined by the immediately adjacent cells. For certain applications, however, some action-at-a-distance must be considered and therefore a larger neighbourhood should be defined (Batty et al., 1997).

The updating of pixels takes place in discrete time steps according to fixed *rules* (Wolfram, 1986). Each time step is called a generation, iteration or cycle. Although these rules can be of very simple construction, CAs produce complex behaviour in most cases. Phipps & Langlois (1997, p. 195) state that 'the CA approach demonstrates clearly that unpredictable structures often emerge from the interplay and spatial propagation of local events even though such events are relatively simple'. The following figure shows a very simple example of a CA:

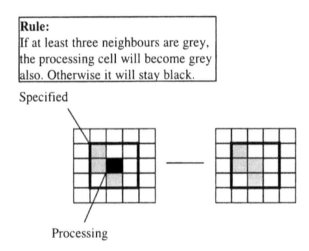

Specified

Processing

Figure 1. *A simple example of a Cellular Automaton*

The CA theory was developed by the physicist and mathematician John von Neuman in the early fifties (von Neumann, 1966). Most following work on CA was of pure physical and mathematical nature as well. It was not until the early eighties that the use of CA in environmental issues was introduced. Since then, CAs have been increasingly applied to problems where there is need to model spatial dynamics. Examples include the modelling of diffusion or growth processes, simulation of evolving patterns, forest fire simulation, land use change and vegetation dynamics modelling (Batty & Xie, 1997; Wagner, 1996). Especially in the field of forest fire simulation, several promising attempts have been made to simulate the spread of fire using Cellular Automata and GIS (Goncalves & Diogo, 1994).

3.   Model Structure

The implementation of the plant dispersal model was strongly based on a conceptual model developed by Cole (1999) at the University of Auckland. The conceptual model describes the important life stages of the plant *Rhamnus Alaternus* as well as the possible cell states and transition rules. Although the conceptual model was adopted, several assumptions and simplifications were necessary to realise the implementation. The general workflow is shown in the following figure.

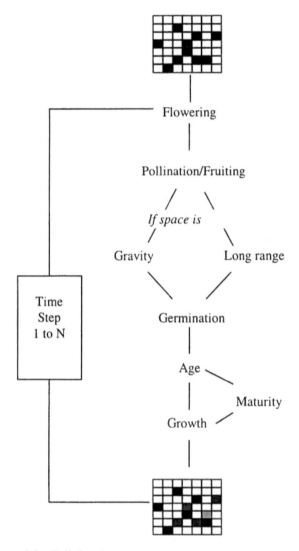

Figure 2. *Flowchart of the Cellular Automaton*

The initial distribution of *Rhamnus* plants represents the starting point of the model. To explore a wider variety of scenarios, both real and random *Rhamnus* distributions were used. In the next step, all adult female plants in the grid are turned into cells with the cell state 'flowering' plants. Flowering plants can be pollinated and therefore fruit if a male plant is within a specified pollinating distance. Flowering cells for which this is true are turned into 'pollinated'/'fruiting' cells. These cells are able to disperse into the neighbourhood if there is space available, i.e. if there are cells with the state 'empty' in their neighbourhood. Dispersal takes place both by gravity, i.e. by falling fruit, and bird-induced with a much wider dispersal range. Cells where dispersal takes place are turned into cells with the state 'germination' and represent juvenile plants. Every inhabited cell grows depending on its age. As long as the new plants are not mature yet, they grow according to the juvenile growth factor. Mature plants grow faster according to the adult growth factor. Juvenile plants reach maturity after a user-specified number of years. Plants that reach a certain size expand into neighbouring cells and turn the state of these cells into their own state value. This concludes the cycle, which represents one year within the life of the plants. The following model runs use the output grids of the last cycle or iteration as input grids. The result of the model runs can be seen as evolving spatial patterns over time representing the dispersal of *Rhamnus Alaternus*. The changing cell states form patterns over time which show the spatial dynamics of plant dispersal.

4. Cells, States, Neighbourhood And Rules Of The CA

According to the specifications of the conceptual model, the cell resolution of the grid used for the model was set to 2 metres. In order to shorten access and calculation time, the extent of the grid was limited to 40 000 m^2

A list of all possible cell states is presented below. The numbers in brackets represent the coding numbers for every state
- Empty cell (0)
- *Germinating plants*
    - Female germinating (1010)
    - Male germinating (1100)
- *Juvenile plants*
    - Female juvenile (2010)
    - Male juvenile (2100)
- *Adult plants*
    Female adult (10)
    - Female flowering adult (15)
    - Female fruiting/pollinated adult (20)
    - Male adult (100)

The size of the neighbourhood, i.e. the number of cells considered for the updating of the processing cell is user-defined (f. ex. 3 x 3, 5 x 5 etc). This enables the analysis of the effects of various sizes of neighbourhoods on the dispersal of *Rhamnus*.

The transition rules incorporate information from static input grids, for instance the availability of perches, as well as the state of the processing cell and its neighbours. Cell attributes that change within the course of the program as for example the age of a plant represent another important model input.

A simple example of a transition rule: A female flowering plant is pollinated if a male plant is in the defined extent of the neighbourhood. The old state of the cell 'female flowering adult' is turned into the new state 'female fruiting/pollinated adult'.

5. Implementation Of The CA In Arc/Info Grid

This chapter describes the main implementation steps of the CA. The core task of the implementation of the Cellular Automaton was to find ways of updating cell states according to the state of neighbouring cells, cell attributes and other input parameters. The transition rules defined in the conceptual model had to be transferred into the ArcInfo/Grid programming language.

ARC/INFO Grid and its programming language, which is based on Tomlin's Map Algebra (1990), offers various pre-defined as well as user-defined functions. These functions consider either single cells (local functions), a group of cells distributed over the grid (zonal functions) or a cluster of neighbouring cells (focal functions). The neighbourhood can be of any shape and size. To model the spatial dynamics of plant dispersal, the focal functions are most important as they enable the consideration of the neighbourhood by performing different mathematical operations on a specified neighbourhood and writing the result to the center cell. Focal functions include simple operations like finding a minimum or maximum value within a specified neighbourhood as well as more complex operations as watershed analysis. The coding of pre-defined focal functions within Grid is easy and straightforward. Most user-defined functions, however, require more complex coding. These functions are created using either the DOCELL block or conditional expressions. These Grid specific commands allow any algebraic or logical expression to be performed on a cell-by-cell processing basis. Masking grids can be used to restrict these operations to a defined neighbourhood.

The coding of the main implementation steps of the CA will be discussed below (see figure 2). To enable an exploratory analysis of *Rhamnus* dispersal, input parameters like the size of the neighbourhood have been kept variable where possible. The different initial distributions of *Rhamnus* plants represent the starting point of the model as far as plant population is concerned. CAPITALS represent Grid commands. Words in % % represent variables.

The complete code can be requested from the author.

5.1 Flowering

All adult females in the grid are turned into flowering females by simply recoding their cell state from '10' to '15'.

5.2 Pollination and fruiting

As females can only fruit if a male cell is within pollinating distance, the specified neighbourhood of every flowering female is checked for a male plant. The neighbourhood analysis is achieved by the focal function FOCALMAX. This function finds the maximum value of the specified neighbourhood for every cell location of the input grid and assigns it to the corresponding cell location in an output grid. This grid has the value 100 for every female cell with a male plant in the specified neighbourhood. The state of each of these female cells is changed from the value '15' for flowering to the value '20' for pollinated/fruiting cell.

5.3 Gravity dispersal

As specified in the conceptual model, fruit disperses either by means of gravity or spreading through birds. Gravity dispersal only affects the local neighbourhood of the fruiting plant. Bird induced dispersal, however, can have long range effects. To implement gravity dispersal in the CA, the dispersing plants must expand into the neighbourhood. One of the simplifications of the model is the random location of dispersal within the specified neighbourhood. Therefore, a mechanism has to be found to first expand fruiting cells into all pixels that represent possible new locations. Then a random cell has to be determined as the pixel where a new plant germinates[1].

The expansion of the fruiting cells into the neighbourhood was accomplished with the EXPAND function. This grid function expands the selected zones (which are all cells of one value, i.e. '15' in this case) by a specified number of cells. The expanded cells all have the same value as the original cell. However, cells cannot expand into already inhabited cells. Therefore, a mask has to be set that masks out all existing pixels by setting them to NODATA and hence prohibit expansion. The concept of NODATA is an important feature of GRID. If inadequate information exists for a cell, the cell location is assigned a NODATA value and the cell is skipped during processing.

If space for dispersal is not available, i.e. existing cells occupy all cells into which the fruiting cells expand, the grid containing the mask will be NODATA. As a grid with only NODATA values that is used in the further course of the CA will lead to an error in the program, a routine has to be added to prevent this. Unfortunately, the implementation of &GOTO statements within cell-by-cell processing is not possible. Therefore, it is not straightforward to skip certain processing steps of a routine. To circumvent this problem, the command ISNULL is applied, which returns the value 1 if a grid has NODATA values. A counting scalar variable was initiated in cell-by-cell processing. This variable adds all values in the grid, which are either 0 for NODATA or 1 for all other values. If the variable equals 0 after adding all values of the grid, all values must be NODATA. The scalar variable from the cell-by-cell processing is written to a regular AML variable, which then allows the execution of the &GOTO statement. Hence, if no space for dispersal is available, the next processing step can be skipped.

If enough space is available for dispersal, the unmasked cells obtain a random value. This makes it possible to select a random cell location within the neighbourhood of the fruiting cell. However, to pick the maximum value of the random numbers in the neighbourhood of every fruiting cell and assign a specific value to that pixel is not straightforward. How can the location of the maximum cell within every neighbourhood be determined? A focal maximum function assigns the maximum value of every neighbourhood to an output grid, on the condition that the centre pixel is a fruiting female. This output grid is then processed with the expand function. A cell-by-cell equality test on the result of these two operations reveals not only the value of the maximum random number in the neighbourhood, but also its location. The following figure demonstrates this procedure.

---

[1] Seeding is not considered in the model. It is assumed that every fruit seeds successfully and therefore new plants germinate.

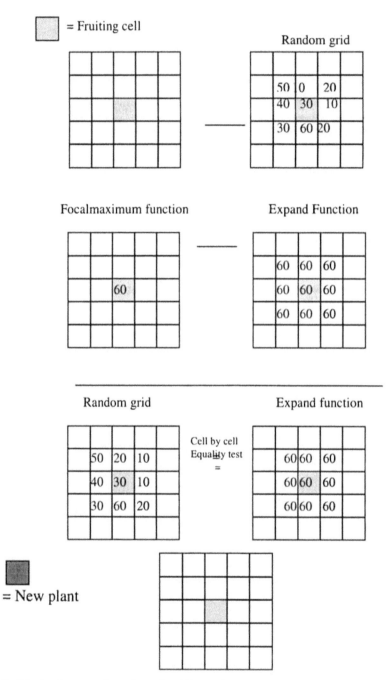

Figure 3. *The implementation of gravity dispersal*

5.4 Long range dispersal

Besides the impact of gravity, fruit are also dispersed by birds moving from existing *Rhamnus* plants to other areas. Contrary to gravity dispersal that has only a local impact on dispersal of *Rhamnus* plants, the long-range dispersal of fruit by birds can spread the plant over a large area. This can be crucial with respect to the invasion of a plant into existing plant ecosystems.

In spite of numerous attempts to model the movement of birds and therefore the dispersal of fruits by birds, the mechanisms of long-range dispersal are still to a large extent unknown. Therefore, it is impossible to model the long-range dispersal on the basis of set rules. The only parameter included in the modelling of bird dispersal is the influence of perches on the birds.

There is a positive correlation between proximity to perch trees and clusters of *Rhamnus Alaternus*. Birds use the perch trees as resting places and have potential to disperse the fruit to the area around the tree. Therefore, a grid with the distance of every cell to the next perch is used in order to define areas where long-range dispersal could take place. The user can determine how often long-range dispersal occurs. Furthermore, a user-defined variable sets a threshold for the distance from the perches. The modulus function is applied to determine whether long-range dispersal takes place in the current cycle or not. The number of the current iteration is divided by the value for the repetition rate for long-range dispersal. If the grid created with the modulus function is zero, dispersal occurs.

5.5 Germination

The life stage 'Germination' assigns the gender of the new plants with the help of random functions as well as user defined distributions and adds the new plants to the already existing ones. Female germinating plants obtain the state '1010', male germinating plants the state '1100'.

5.6 Age

The subroutine 'Age' is an important part of the CA as it represents the mean of keeping track of the age of plants. Age determines whether plants are mature or juvenile, which influences the growth rate and ability to reproduce.

In the first cycle, the grid startage is created, which contains the original age of all existing plants. Existing plants are assigned the age 20 (which is arbitrarily chosen), while all other cells (even empty cells) obtain the value 1. This is necessary in order to compare the recent age of every plant with the age of the plant when the model started. The variable %.index% which counts the number of iterations of the CA is set to a scalar variable, in order to use the variable within a DOCELL block. In the first cycle, all adult female and male *Rhamnus* plants obtain the new value age1 = startage + 1. For the first cycle, the grid called age%.prev_index% is set to startage - 1 for all cells containing plants. For empty cells, the grid is assigned the value 0. The variable %.prev_index% states the number of the previous iteration. A grid that records the age of the previous cycle for every cell is created.

For every further cycle, adult plants age one year per cycle and obtain the new age age%.index% = age%.prev_index% + 1. Germinating plants are assigned the value 1. As germinating plants will be turned into juvenile plants at the end of every cycle, it is not necessary to consider the previous age of these plants. Juvenile plants obtain the value new age age%.index% = age%.prev_index% + 1.

5.7 Growth and Maturity

After the age of every plant is updated, juvenile pixels are first evaluated whether they are old enough to mature. According to the user input for the maturing age, juvenile cells mature at the age of %.maturity% + 1 and obtain the state '10' for female adult or '100' for male adult.

As long as the plants are juvenile, the juvenile growth rate applies. Therefore, if age%.index% which represents the absolute present age of every plant is lower than 6, growth is calculated as Juvenile growth rate (0.125 m2/y) * age%.index%.
For adult plants originating from the initial distribution, growth is calculated as Adult growth rate (0.25 m^2/y) * (age%.index% - startage).

The present age has to be subtracted from the original age in order to determine the number of years where growth has taken place. For adult plants that germinated and matured in the course of the CA, growth is determined by
0.25 * (age%.index% -5)
as the years where the plant grew as a juvenile plant cannot be considered for adult growth.

As the cell resolution is 2 metres, the plant has to grow 4 m^2 to occupy a whole new cell. It is assumed that new cells are populated only when a plant occupies a new cell to the full extent. Partially occupied cells are not considered. The MODULUS function is applied to determine if growth reaches a multiple value of 4. If growth is a multiple of 4, the result will be 0. In this case, the processing cell grows into one of the neighbouring cells. The location of that cell is determined in the same way as shown for the germinating cells. The command EXPAND expands the growing cell into the neighbourhood, a mask is set to prevent the cell to expand into existing cells and a random function assigns random values to all expanding cells. The location of the cell with the maximum random number will be the location of the grown cell. The newly inhabited cells are added to the existing plant distribution. After the determination of growth, germinating plants are turned into juvenile plants. All other cells remain unchanged.

5.8 Temporal dynamics

The temporal dimension can easily be simulated by creating different output grids for every time step and using these grids as input for the next iteration. One time step represents one year in the life cycle of *Rhamnus Alaternus*. The cycle is iterated as many times as specified by the user.

6. Results

As the aim of the model was to deliver an exploratory tool to analyse the effects of different parameters on the dispersal of plants, input values have been kept variable were possible. The input grid and therefore the initial distribution of plants, the number of iterations, the size of the neighbourhood, the time interval between long range dispersals, the age at which juvenile plants mature and the distance from perches were varied for different model runs. Figures 4 and 5 show the evolving pattern for the spreading of *Rhamnus Alaternus* in dependence on different input values.

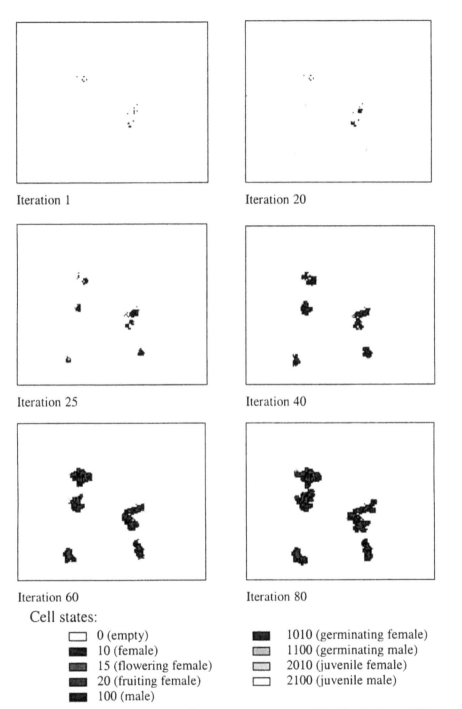

Iteration 1

Iteration 20

Iteration 25

Iteration 40

Iteration 60

Iteration 80

Cell states:

☐ 0 (empty)	■ 1010 (germinating female)
■ 10 (female)	▨ 1100 (germinating male)
▦ 15 (flowering female)	▧ 2010 (juvenile female)
■ 20 (fruiting female)	☐ 2100 (juvenile male)
■ 100 (male)	

Figure 4. *Initial distribution and evolving pattern after 10, 25, 40, 60 and 80 iterations. Input parameters: initial distribution: 35 % female plants, 80 iterations, 3 x 3 neighbourhood, 5 years time interval between long range dispersal, juvenile plants mature after 5 years, distance from perches: 4 m*

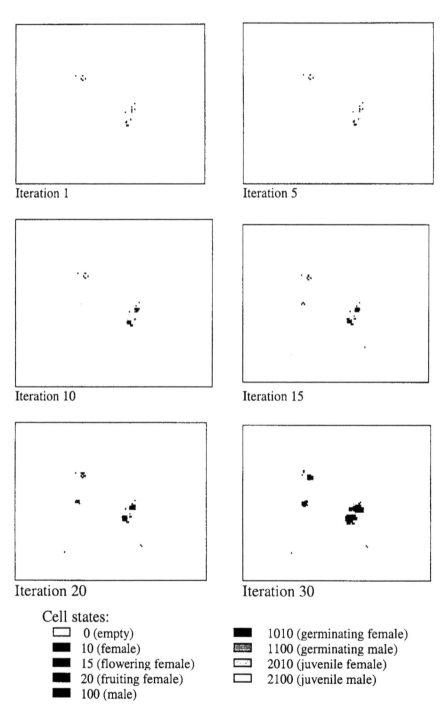

Figure 5. *Initial distribution and evolving pattern after 5, 10, 15, 20 and 30 iterations. Input parameters: initial distribution: 50 % female plants, 30 iterations, 3 x 3 neighbourhood, 7 years time interval between long range dispersals, juvenile plants mature after 5 years, distance from perches: 2 m*

Existing cells continuously spread into the neighbourhood. No particular pattern evolves, the existing clusters grow according to the transition rules and states of the cells. As the direction of dispersal is random, growth is equally distributed around the cluster centres. At iteration 10, new clusters develop in the lower left part of the grid due to bird induced long range dispersal. At first, these new clusters grow and spread fast, because there is more space available for dispersal as in the already existing clusters. After around forty years, the original clusters grow faster, while the growth of the new clusters finally comes to an end around iteration 80. All female cells of the lower right cluster, for instance, are surrounded by male cells, which stops the evolution until new long range dispersal takes place.

Generally, dispersal is slower as long range dispersal occurs every 7 years only. Furthermore, growth is more spatially confined. As only cells are considered, which lie within a 2 m distance from a perch, fewer suitable cells are available and therefore long range dispersal is more restricted compared to the first example run. New clusters do develop but they remain very confined and small.

7. Discussion And Conclusion

The aim of the work presented in this paper was to implement a Cellular Automaton approach within a GIS to enable truly dynamic modelling. The results show that ARC/INFO Grid allows an implementation of a Cellular Automaton, which delivers an exploratory approach to model plant dispersal.

Though being a simplified model, the CA behaves differently for different input parameters and therefore proves its capability of observing the effects of changing input parameters on the dispersal of *Rhamnus*. The developed model needs only few input parameters which can easily be gathered through field work or literature study. The CA will then be fed with a range of values for every input parameter. The results of these model runs enable the ecologist to estimate the spreading of *Rhamnus* plants in respect to the number of bird perch trees for instance. Climatic factors like continentality of the area of interest or the occurrence of warm coastal breezes are important as well and can easily be added to the model as additional GIS layers.

In spite of the successful implementation, several problems have been encountered using Grid making coding more complex and performance less efficient. As discussed above, Grid does not allow easy determination of a certain value as well as the location of this value in the grid. Complex coding is required, which leads to lower efficiency of the program. Conditional statements in Grid can be done on a cell-by-cell basis. This is a very helpful tool for modelling, as the state of a cell can be changed if other input grids or variables have a certain value. On the other hand, the output grid is created even if the conditional expression is false for all cells of a grid. This can be annoying as grids without information will be assigned NODATA and can lead to errors in the course of the program.

The use of look-up-tables as proposed by Wagner (1997) was not appropriate as the information needed had to be drawn from different grid layers. The possibilities to work with attribute tables in Grid are very limited. Therefore, a very high number of grids must be created as Grid lacks functionality in updating grids. This leads to inefficient renaming and killing of grids and to an unnecessarily high number of grids, which results in slower processing time.

Regardless of these limitations, ARC/INFO Grid proved to be a tool to simulate CA behaviour in a Geographical Information System. The two main features were the neighbourhood analysis and the possibility to perform conditional statements on a grid, multiple grids, scalars, numbers or a combination of all. The GRID neighbourhood functions simulated the concept of proximal space whereas the GRID conditional statements were mainly used to code the transition rules of the CA.

An evaluation of the model is difficult, as there is not enough field data available yet to be able to compare the model results with real world data. Many assumptions had to be made in the course of the implementation. Every assumption simplifies the real world and therefore makes the model less realistic and accurate. However, the aim of the model is not to simulate the spreading of *Rhamnus* in its whole complexity and to achieve an exact representation of the real world, but to have an explanatory approach where we can play with different input values and explore the effects of a change in input on the spreading of *Rhamnus Alaternus*.

The implementation of further parameters would certainly improve the performance of the model. As far as the computational environment for modelling plant dispersal is concerned, Cellular Automata Machines would represent an alternative possibility to develop a model incorporating a CA. A comparison with the implementation in a Cellular Automata Machine would illustrate the rather complicated implementation procedure in a GIS delivers a modelling tool which performs as well as the very application specific Cellular Automata Machine.

References

Batty, M., Couclelis, H. et al. (1997): Editorial, *Environment and Planning B: Planning and Design* **24**.

Batty, M. & Xie, Y. (1994): From cells to cities. In: *Environment and Planning B: Planning and Design* **21**, s31-s48.

Benati, S. (1997): A cellular automaton for the simulation of competitive location. In: *Environment and Planning B: Planning and Design* **24**, p. 205-218.

Buckley, A. (1994): *Cellular automata theory applied in simulation of forest change.* Poster presented at the 90th annual meeting of the Association of American Geographers, San Francisco, CA.

Burrough, P. A. and R. A. McDonnell (1998): *Principles of Geographical Information Systems*. Oxford, Oxford University Press.

Cole, V. (1999): *Modelling the invasion of Rhamnus Alaternus in GIS based Cellular Automata*, Master Thesis, Geography Department, University of Auckland, New Zealand.

Couclelis, H. (1991): Requirements for planning relevant GIS: a spatial perspective. In: *Papers in Regional Science* **70**(1), p. 9-19.

Couclelis, H. (1997): From cellular automata to urban models: new principles for model development and implementation. In: *Environment and Planning B: Planning and Design* **24**, p. 165-174.

Environmental Systems Research Institute (ESRI) (1991*): Cell-based Modeling with GRID.* Redlands, California.

Goncalves, P. P. & Diogo, P. M. (1994): *Geographic Information Systems and Cellular Automata: a new approach to Forest Fire Simulation.*

Hogeweg, P. (1988): Cellular Automata as a Paradigm for Ecological Modelling. In: *Applied Mathematics and Computation* **27** (1), p. 81-100.

Peuquet, D. J. (1994): It's about time: a conceptual framework for the representation of temporal dynamics in geographic information systems. In: *Annals of the Association of American Geographers* **84**, p. 441-461.

Phipps, M. & Langlois, A. (1997): Spatial dynamics, cellular automata, and parallel processing computers. In: *Environment and Planning B: Planning and Design* **24**, p. 193-204

Preston, K. & Duff, M. J. B. (1984*): Modern Cellular Automata Theory and Applications.* New York, Plenum Press.

Silvertown, J., S. Holtier, et al. (1992): Cellular automaton models of interspecific competition for space - the effect of pattern on process. *Journal of Ecology* **80**, p. 527-534.

Tobler, W. (1975): Cellular Geography. In: *Philosophy in Geography*, p. 379-386.

Toffoli, T. & Margolus N. (1987): *Cellular Automata Machines - A new environment for modeling.* Cambridge, Massachusetts, London, England, MIT Press.

Tomlin, D. C. (1990): *Geographic Information Systems and Cartographic Modeling*, Englewood Cliffs, N. J., Prentice Hall.

von Neumann, J. (1966): *Theory of Self-Reproducing Automata.* Champaign, Illinois, University of Illinois Press.

Wagner, D. F. (1996): Cellular Automata as Analytical Engines in Geographic Information Systems, *1st International Conference on GeoComputation*, Leeds, p. 670-688.

Wagner, D. F. (1997): Cellular Automata and geographic information systems. In: *Environment and Planning B: Planning and Design* **24**, p. 219-234

White, R. & Engelen, G. (1997). Cellular automata as the basis of integrated dynamic regional modelling. *Environment and Planning B: Planning and Design* **24**: 235-246.

Wolfram, S. (1986): *Theory and Applications of Cellular Automata*, Singapore, World Scientific Publishing Co Pte Ltd.

Wu, F. (1998): SimLand: a prototype to simulate land conversion through the integrated GIS and CA with AHP-derived transition rules. *International Journal of Geographical Information Science* **12**(1), p. 63-82.

11

Preliminary Analyses Results Of Forest Plant Diversity And Distribution On Mt. Medvednica, Croatia

Sven D. Jelaska

This paper reports on the results of analyses of 28 forest vegetation plots (50 x 50 meters) situated along a transverse transect in Medvednica Nature Park, Croatia. The purpose of the research was to establish a correlation between the plant diversity and environmental factors with an emphasis on concentration of chemical elements in the topsoil. Models were developed to demonstrate a possible approach for defining priority areas for the protection of plant diversity with the use of GIS. Multiple linear regression was used for the development of a predictive model of the number of plant species, with soil acidity and concentrations of cobalt, chromium, magnesium and yttrium as independent variables. For rare species, logistic regression and equal weight predictive models were developed using the altitude, soil acidity and concentrations of nickel, cobalt and potassium extracted as independent variables by canonical correspondence analysis (CCA). Developed models were incorporated in GIS, and areas with highest plant diversity and highest number of rare plant species were defined. Based on the results of the cluster analysis of the species composition, a dataset of vegetation type was derived and used for selection of an additional area in order to protect the complete plant diversity of the surveyed area. Appropriateness of selected statistical methods were discussed, as well as advantages of GIS-based approach in analyses of plant diversity and distribution, and selection of areas important for preservation of plant diversity.

1 Introduction

Completion of the mapping of Croatian flora should be one of the primary goals in botanically related activities in the country. However the results achieved to date indicate that with the traditional methods of mapping and the limited number of active botanists, this goal is highly unlikely to be accomplished in near future. The use of Geographic Information Systems (GIS), in combination with statistical models could provide an alternative approach to producing information nationwide, with the production of an atlas of the probable distribution of Croatian flora.

As a part of a trial mapping of the Medvednica Nature Park[1], based on a standard proposed for the mapping of Croatian flora (Nikolić et al., 1998) 28 forest vegetation plots were surveyed. This demonstrated a methodology for the analysis of flora distribution and predictive mapping using GIS and modelling in the management of protected areas. This sample size was determined by the availability of time and resources.

Mean annual precipitation and temperature have been used to explain and model plant distribution by Kadmon & Danin (1997) and Austin (1998), however, in this research other abiotic factors have been used similar to those of McCune & Allen (1985) and Kazda (1995). Because of the vicinity of the Nature Park to the city of Zagreb, the biggest industrial centre in the country, the concentrations of base metals and other chemical elements in the topsoil were studied with respect to the distribution of the flora in the Nature Park.

The paper describes the data collected and the statistical and GIS analyses used, followed by an interpretation of the results.

2   Study Area

The Medvednica Nature Park occupies the western part of Mt. Medvednica, which is situated north of Zagreb, the capital of Croatia, with its approximate centre at 45°55' N latitude and 16° E longitude (Figure 1A). Temperate climatic conditions are dominant, with associated vegetation types such as oak, beech and beech-fir forests, although on some south-facing parts there are thermophilic floristic elements. This paper describes part of a survey involving fieldwork over the entire nature park area of approximately 22000 ha divided into 135 MTB 1/64 grid units, as described by Nikolić et al. (1998).

3   Methods

3.1. Field survey

Three types of data were gathered directly in the field. Geographical position of the plots surveyed, geochemistry data including soil acidity, and floral assemblage were collected along the central Gračani - Donja Stubica transverse profile, determined by nine MTB 1/64 grid units. A total of 28 rectangular 50 x 50 metre plots of forest vegetation were defined, in an altitudinal gradient from 195 to 954 meters above sea level (Figure 1B). The size of the plots was chosen to ensure that majority of species would be recorded even in the acid beech forest, that has smaller species diversity. In two plots (No. 17 and 28) the requirement of a regular rectangular shape was rejected in order to retain uniformity in terrain features and vegetation cover. The main criteria for selection of plot positions were vegetation type, uniformity of relief, relative closeness to the forest roads and hiking tracks to facilitate the access to the plots. The plots were defined with a measuring tape and compass, and marginal trees marked with paint.

---

[1] "Nature Park" – similar to "National Park" but less restricted in terms of the permitted human activities in its area.

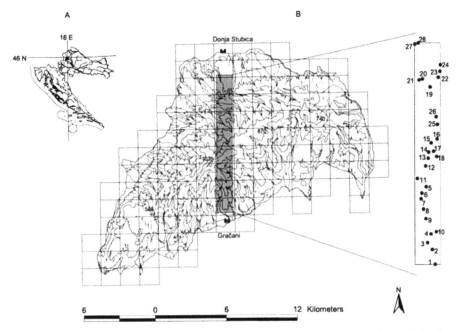

Figure 1. *A - a geographical position of Medvednica Nature Park, **B** - Medvednica Nature Park with the MTB 1/64 grid, position of the 28 forest plots and an area enclosed by interpolation of data from survey plots.*

The geographical position of the corners of the 28 plots were surveyed using a Trimble Geoexplorer II GPS (Global Positioning System) receiver in the field, and a Trimble 4600 LS base station at a known location for post-processing differential correction using Pathfinder office software (Trimble Navigation Ltd.).

For geochemical analyses, samples of the upper 10 cm of the soil were collected, between October 22nd and November 6th, 1997, after the removal of dead organic material from the surface. For each plot, 8 samples with a 12.26 cm^3 volume cylinder to a depth of 10 cm were collected, taken at 10 m intervals in a diagonal across the plot, starting and ending at opposite corners of the plot (Scholz et al., 1994). All of the samples were combined for each plot, dried in the air at room temperature and sent to ACME Analytical Laboratories Ltd., Vancouver, Canada for 1E ICP-ES (Inductively Coupled Plasma Emission Spectrometry) analyses of 35 elements and pH analyses (with Orion meter using temperature adjusted pH electrode).

The presence or absence of plant species was observed during the 1998 growing season, at monthly intervals from February to September, to ensure the complete recording of the flora of the plots. All floral assemblages were classified in a standard manner using the Index Florae Croaticae (Nikolić 1994, 1997, 2000).

3.2. Creation of the GIS for the study area

Topographic data from a digital 1:25000 scale map were obtained from Croatian Radio-Television for the complete area of the Nature Park. For the surveyed area other paper map data were also available e.g. vegetation, however these data were not used in the analysis because of their nominal scale values and the relatively small sample (i.e. 28 plots).

The surveyed plots were compiled into polygon datasets, and used to derive the elevation of the centre of the plot by overlaying the polygons with a 20-meter cell Digital Elevation Model (DEM), produced from the 1:25000 digital topographic map. Mean slope within the plots was also derived from this DEM, using the sum of the slope value multiplied by the percent of plot area that grid cell, or its part, occupied. No data was used for aspect.

A raster thematic layer was created, using the IDW (Inverse Distance Weighting) method (e.g. Preston et al., 1996), for the measured geochemical variables that were expected to be of significance to the distribution of the plant species. Interpolated area enclosed by the surveyed plots covered approximately 1500 ha. The cell size of 50 x 50 meters was determined by the size of the surveyed plots, since the number of present plant species is directly dependent on the size of an area (larger surface - more plant species and vice versa). Therefore cell size had to be the same as that of the sample plots.

These layers were used as inputs to the development of models, and the subsequent projection of the predicted number of plant species and species distribution modelled on the area surrounding the survey transect. Interpolations, spatial overlapping of thematic layers, map editing and all other GIS-based tasks were performed using ARC/INFO© 7.0.4, ARCGrid© and ArcView© software (ESRI, Inc.).

3.3. Statistical analyses

The estimation of plant diversity, expressed as the number of plant species in the surveyed area, used the multiple regression model developed by Jelaska & Nikolić, (2000), and analysis of the floristic composition of the plots used the Canonical Correspondence Analysis (CCA) inside the CANOCO program (Ter Braak, 1988; 1990). Geochemical variables that had element concentrations below the detection level on at least one plot were excluded from further analyses. Nominal variables were excluded from the analyses, although CCA can deal with them, because the size of sample, i.e. 28 plots, was too small for the number of variables, which would increase significantly when dummy variables were used (a necessary step when using nominal variables).

From 18 environmental variables included in the CCA analyses, a further four were excluded, due to their high Variance Inflation Factor (VIF). The values of the remaining 14 environmental variables, the number of recorded plant species and the presence of modelled plant species on each plot surveyed are shown in Table 1 (with the addition of chromium that was included in the plant species number regression model).

The variables that best explained the first four canonical axes were used for the development of predictive occurrence models for the species in Table 2.

For the *Platanthera bifolia* the multiple logistic regression (logit) model of spatial distribution was developed in the STATISTICA 5.0 program package (StatSoft, 1995). The final model was selected based upon the highest accuracy of model tested on input data and the level of significance of the coefficients (based on the chi-square test). The variance in the remaining five species was too high for multiple logistic regression models to be developed, therefore, an equal weight model was developed using the overlapping of values from the range of environmental variables extracted by the CCA analysis, from each plot.

The amount of species variation explained by the first four ordination axes was used for determining the priority of inclusion of the environmental variables in the model, where combinations of environmental variables that were most accurate in the input data were chosen for use in the final models.

Vari-able	Cu (mg/kg)	Pb (mg/kg)	Ni (mg/kg)	Co (mg/kg)	Mn (mg/kg)	Ca (%)	P (%)	Cr (mg/kg)	Mg (%)	Al (%)	K (%)	Y (mg/kg)	pH	Alti-ude (m a.s.l)	slope (deg)	A	B	C	D	E	F	Species No
Det-ection	2 mg/kg	5 mg/kg	2 mg/kg	2 mg/kg	5 mg/kg	0.01%	0.002%	2 mg/kg	0.01%	0.01%	0.01%	2 mg/kg										
Plot 1	262	23	84	28	1252	1.00	0.059	141	1.09	7.48	0.82	24	4.7	389	12.7	1	0	0	1	1	0	62
Plot 2	356	28	73	17	1411	0.40	0.053	70	0.43	5.79	1.11	20	4.9	493	23.1	1	0	0	1	1	0	50
Plot 3	101	45	30	5	212	0.10	0.067	60	0.40	5.52	1.13	7	3.8	531	20.2	0	0	0	0	0	0	19
Plot 4	324	230	65	23	6302	0.50	0.147	62	0.62	5.76	0.87	13	5.4	647	11.3	1	0	0	0	0	0	41
Plot 5	152	40	44	29	1726	2.26	0.104	108	1.24	7.48	1.03	32	4.7	825	23.5	0	0	0	0	1	0	54
Plot 6	119	17	35	29	1696	2.13	0.084	75	1.56	7.17	0.59	38	5.2	766	28.2	0	0	0	0	0	0	50
Plot 7	76	53	29	14	576	2.07	0.062	86	1.53	4.85	0.31	17	4.0	768	30.4	0	0	0	0	1	0	19
Plot 8	66	47	33	13	499	2.35	0.067	94	1.48	4.98	0.35	16	3.9	774	20.8	1	0	0	0	0	0	38
Plot 9	63	64	36	14	593	2.55	0.048	110	1.57	6.04	0.37	18	3.8	675	15.3	0	0	0	0	0	0	19
Plot 10	210	30	68	14	519	0.10	0.033	90	1.00	7.98	1.88	8	4.0	530	26.7	0	0	0	0	0	0	16
Plot 11	115	95	53	25	1123	2.11	0.085	115	1.64	6.60	0.72	22	4.2	954	23.4	0	0	0	0	0	0	43
Plot 12	116	48	29	28	1596	2.40	0.099	57	2.42	6.84	0.31	38	4.1	860	29.4	0	0	0	0	0	0	28
Plot 13	144	88	41	19	1862	1.00	0.159	80	1.47	6.76	1.14	19	4.9	827	17.9	0	1	0	0	0	0	36
Plot 14	141	44	57	19	2464	1.35	0.059	62	0.75	6.03	1.56	22	5.7	746	19.4	0	0	0	0	0	0	48
Plot 15	225	44	57	21	1603	0.40	0.034	84	0.50	5.96	1.04	17	4.9	709	11.1	0	0	0	0	1	1	54
Plot 16	133	35	171	34	1298	2.45	0.041	369	3.15	6.18	0.70	10	5.1	678	14.1	0	0	0	0	1	1	51
Plot 17	155	51	42	13	969	0.15	0.051	68	0.47	5.93	1.19	10	4.1	744	15.8	0	0	0	0	0	1	47
Plot 18	225	54	39	9	436	0.19	0.057	75	0.37	5.74	1.19	7	4.0	840	18.3	0	0	0	1	0	1	25
Plot 19	521	50	46	22	1720	0.44	0.064	85	0.65	7.17	1.29	12	4.1	375	16.7	0	0	0	0	1	0	39
Plot 20	250	41	27	13	1063	0.36	0.040	66	0.59	6.15	1.38	9	4.4	303	11.6	0	0	0	0	0	0	25
Plot 21	81	45	17	5	199	0.23	0.044	44	0.31	3.95	0.92	6	3.8	280	12.9	0	0	0	0	0	0	20
Plot 22	131	36	41	31	2117	0.93	0.060	94	0.89	7.27	0.93	18	4.8	366	16.6	0	0	0	1	1	1	53
Plot 23	216	38	46	30	1326	0.57	0.065	88	0.73	6.93	0.95	16	4.5	275	13.8	0	0	1	0	0	1	49
Plot 24	75	38	24	10	517	0.26	0.041	62	0.58	5.72	1.36	9	4.0	279	17.2	0	0	0	0	0	0	27
Plot 25	130	51	50	14	826	0.28	0.048	96	0.60	6.43	1.52	9	4.7	604	17.0	0	0	0	0	1	1	35
Plot 26	168	53	29	14	2035	0.24	0.056	58	0.45	5.58	1.19	10	4.5	617	12.5	0	0	0	0	1	1	39
Plot 27	148	36	26	11	654	0.27	0.035	65	0.50	5.84	1.38	10	4.0	245	5.3	0	0	0	0	0	0	26
Plot 28	195	28	37	13	821	0.78	0.063	69	0.75	6.42	1.58	16	5.1	195	11.6	1	0	0	0	0	0	44

Table 1. *Species Observations*: A : *Erythronium*; B : *Lilium*; C : *Ceph. Dam*; D : *Ceph. Long*; E : *Platanthera*; F : *Daphne*

Latin name	Family	state and protection
Erythronium dens-canis L.	*Liliaceae*	vulnerable, not protected
Lilium martagon L.	*Liliaceae*	endangered, protected
Cephalanthera damasonium (Mill.) Druce	*Orchidaceae*	endangered, protected
Cephalanthera longifolia (L.) Fritsch	*Orchidaceae*	endangered, protected
Platanthera bifolia (L.) Rich.	*Orchidaceae*	vulnerable, protected
Daphne laureola L.	*Thymelaeaceae*	rare, protected

Table 2. *Plant species whose distributions were modelled with theirs endangered state and status protection according to Croatian law.*

The floral assemblage of the survey plots was analysed using cluster analysis, to detect the main vegetation types present. Since the data about plant species were binary (present/absent) a complete linkage agglomerative cluster algorithm was used with "simple matching distance" index in SYN-TAX-pc 5.0 program (Podani, 1993) to produce a dendrogram.

3.4. GIS analyses

A cell-based analysis of the environmental data layers was used to project the predicted number of plant species present for each cell, across an area enclosed by IDW interpolation, by entering the interpolated values into the derived regression models. Calculated values were then reclassified into three classes following the scheme: **class 1:** 48-60 plant species; **class 2:** 33-47 plant species; **class 3:** 18-32 plant species. A final polygon layer was produced by dissolving the boundaries of the layers based upon the class values.

The data for *Platanthera bifolia* was reclasified because the ouput from the logit regression model was grid cell values in a range from 0 to 1, which actually represents the probability of species occurrence. All cells with values smaller then 0.5 were coded to 0 and the others to 1. In this way all the species predicted distribution layers had only two possible values, 0 for unsuitable and 1 for suitable environmental conditions i.e. absence or presence of species. The sum of simultaneous predicted occurrences of species was calculated and reclassified into 4 classes: **class 1:** 5 species; **class 2:** 3 or 4 species; **class 3:** 1 or 2 species; **class 4:** 0 species. The polygon layers of the predicted distribution of all the six modelled species were then spatially joined.

The sum of the classes of diversity and rare species layer was calculated for each newly formed polygon. The level of priority for the protection of an area was inversely proportional to the calculated sum of the two class values described above, and this sum (ranging from 2 to 7) was named Important Plant Diversity Area (IPDA).

Using the values of altitude and soil acidity, that were identified as being the most significant variables for the interpretation of the first two canonical axes, the ranges of each cluster (i.e. vegetation type) in the dendrogram were established, and a map produced of the predicted distribution of vegetation types. Overlapping the "vegetation" map with the IPDA map, the vegetation type not covered by IPDA class 2 and 3 (most representative for the protection of plant diversity and rare and protected species) was identified. Afterwards, using the threshold value of 10 ha, a polygon with the lowest IPDA value (meaning highest plant diversity) was added to a final map of areas that best represent the plant diversity on the transect studied.

4 Results

In total, 216 vascular plant species were recorded on the survey plots.

The estimated number of plant species was calculated by entering the interpolated values of the significant environmental variables into the following equation:

$$Nosp = e^{(-1.993+0.292*Co+0.477*Cr-0.49*Mg+0.399*Y+0.991*pH)}$$

where Nosp is the estimated number of plant species, and Co, Cr, Mg, Y and pH concentrations of cobalt, chromium, magnesium, yttrium and soil acidity multiplied by coefficients derived with stepwise multiple regression (Jelaska & Nikolić, 2000). A cartographic presentation of the reclassified values is shown in Figure 4A.

The results of the CCA analysis with 14 environmental variables are shown in Table 3. Canonical coefficients for first four ordination axes and their T-values are displayed in Table 4, indicating the altitude, soil acidity, concentration of potassium and nickel as most significant for the interpretation of the first four axes, respectively. These four environmental variables, together with concentration of cobalt, which contributed significantly to the interpretation of the first and fourth axis (Table 4), were included in the development of a multiple logistic regression model and equal weight model for plant species occurrences.

An occurrence probability map for the *Platanthera bifolia* was produced entering the interpolated values of environmental variables into the equation:

$$Pocc = 1/(1+e^{-(-11.4708+0.0001*Ni+0.2386*Co+3.0884*K+0.7624*pH)})$$

where Pocc is the probability of *Platanthera* occurrence, and Ni, Co, K and pH concentrations of nickel, cobalt, potassium and soil acidity multiplied by coefficients from the logistic multiple regression model (chi-square value of coefficients $\chi2(4)=14.019$, $p=0.00724$).

The ranges in values of environmental variables included in the equal weight models are shown in Table 5. Values of nickel, cobalt and potassium are multiplied by 100, and that of pH by 10, and rounded up to integer values (necessary for grid to polygon conversion). Since *Cephalanthera damasonium* and *Lilium martagon* were observed on only one plot, their ranges were obtained by subtracting and adding the standard deviation from, or to, the measured value of the variable. An example map of areas of potential occurrence is displayed in Figure 2 for *Daphne laureola*. Figure 4B shows the classified values of the simultaneous occurrences of all six modelled species.

On the cluster analysis dendrogram (Figure 3) four main clusters were formed with two middle ones further divided into two subclasters. The range of values of modelled vegetation types, defined by clusters, are presented in Table 6 (plot No. 4 was added to cluster B2). Since cluster D (Figure 3) is represented by only one plot, the value of the mean standard error was subtracted and added from, or to, the measured value of soil acidity for modelling the spatial distribution of that vegetation type. For cluster A, a selection of suitable cells was refined by the range of cobalt concentrations, to ensure differentiation from the B1 cluster. Joining succession of cells to appertaining vegetation

Axes	1	2	3	4	Total inertia
Eigenvalues :	0.499	0.326	0.270	0.223	4,508
Species-environment correlations :	0.984	0.867	0.957	0.972	
Cumulative percentage variance					
of species data :	11.1	18.3	24.3	29.2	
of species-environment relation :	18.8	31.0	41.2	49.5	
Sum of all unconstrained eigenvalues					4,508
Sum of all canonical eigenvalues					2,660

Table 3

Axes	1	2	3	4	1	2	3	4
	Oc	oc	oc	oc	T	t	t	T
altitude	-1.012	-0.469	-0.232	0.157	-12.772	-1.838	-1.724	1.455
slope	0.208	-0.406	0.101	-0.139	2.607	-1.577	0.741	-1.276
PH	-0.527	1.211	-0.608	-0.144	-3.938	2.807	-2.673	-0.791
Cu	0.040	0.377	-0.300	0.093	0.505	1.470	-2.218	0.863
Pb	-0.112	1.042	0.652	-0.937	-0.660	1.907	2.264	-4.064
Ni	1.027	-0.915	-0.342	-1.079	7.776	-2.150	-1.524	-6.004
Co	-1.174	0.066	0.156	1.150	-7.268	0.126	0.569	5.232
Mn	0.411	-1.554	-0.236	0.993	2.295	-2.695	-0.777	4.079
Ca	0.366	0.123	0.227	-0.043	2.534	0.264	0.926	-0.221
P	0.105	-0.051	-0.200	-0.402	0.995	-0.149	-1.115	-2.799
Mg	-0.691	0.920	0.601	0.369	-4.334	1.792	2.217	1.703
Al	0.383	-0.208	-0.424	-0.799	2.949	-0.497	-1.923	-4.526
K	-0.482	0.506	0.701	0.415	-3.762	1.227	3.224	2.385
Y	0.382	0.318	0.137	-0.748	2.506	0.648	0.531	-3.613

Table 4

species	altitude	pH*10	Ni*100	Co*100	K*100
Erythronium dens-canis	195-774	-	3300-8400	1300-2800	-
Lilium martagon	522-970	51-62	2752-8548	1029-2671	114-196
Cephalanthera damasonium	51-500	40-50	1702-7498	2180-3821	54-136
Cephalanthera longifolia	366-493	-	-	-	82-129
Daphne laureola	275-840	40-49	2900-5700	900-3000	-

Table 5

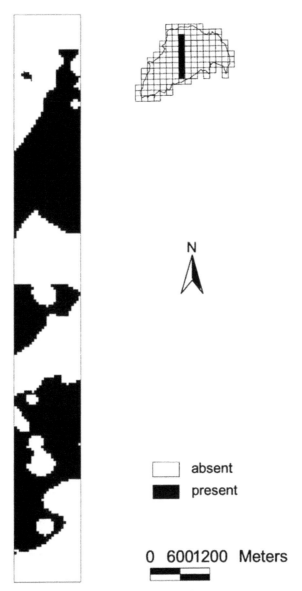

Figure 2. *An example of predicted distribution of Daphne laureola L. (rare, protected).*

types were in the same order as they are in the Table 6, meaning that cells for the cluster D were selected first, then for the cluster A, followed by cluster B1 etc. For the last type (B2 cluster in Figure 3) an additional condition was established stating that none of the previously classified grid cells could be chosen for this last type, and some cells remained unclassified. The final output dataset of the vegetation types is shown in Figure 4D.

The IPDA map is displayed in Figure 4C, and the final map of the areas of most significance for the protection of plant diversity and rare and protected species, with an additional area determined after spatial overlap of IPDA and the vegetation map, is displayed in Figure 4E.

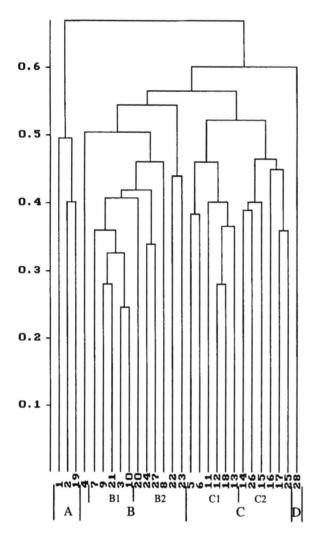

Figure 3. *Dendrogram from cluster analyses of floristic composition of the survey plots using the complete linkage agglomerative cluster algorithm with simple matching distance index. (Cluster A: oak forest; cluster B1: acid beech forest; B2: beech forest; cluster C1: beech-common fir forest; cluster C2: beech-common fir forest; D: common alder forest)*

cluster - vegetation type	altitude	pH*10	Co*100
D - common alder forest: al	<200	50-52	-
A - oak forest: q	375-493	41-49	1700-2800
B1 - acid beech forest: la	280-768	38-40	-
C1 - beech and fir forest: af1	766-954	40-52	-
C2 - beech and fir forest: af2	604-746	45-57	-
B2 - beech forest: lo	275-774	39-54	-

Table 6

5 Discussion

Since the predicted number of present plant species was for a 50 x 50 meters, a relative measure of plant diversity richness was used, where class 1 indicated the highest, and class 3 lowest diversity, independently of the size of an area.

The cumulative percentage variance of species data from the CCA analysis was not very high (Table 3), but that is common in ordination analysis which deal with diverse species data (e.g. Becker et al., 1998; Gottfried et al., 1998). Species data, especially those relating to presence/absence, are very noisy. However, such ordination analyses can also be informative (Ter Braak, 1990) and extract the environmental variables most significant for explaining of variance in the species data.

Gottfried et al., (1998) as well as Guisan et al., (1999) use results of CCA analysis as direct inputs to GIS analyses through the Euclidean distances of each modelled spatial unit in the CCA multidimensional space. In this paper the results of the CCA analysis were only used as a method for the selection of environmental variables that are most significant for explaining spatial variation in plant species distribution. The multiple logistic regression and equal weight model used were significantly less computationally demanding, and more appropriate considering the nature of the input data (small sample with large number of recorded species). For rare species (e.g. *Lilium martagon, Cephalanthera damasonium*) observed on just one plot, the amount of explained spatial variation is very small. Therefore, a significant computational effort would probably not be justified by the increase in the accuracy of the predictive model that would be output.

Since the logit model developed for *Platanthera bifolia* used data from all sites, it was capable of greater accuracy than an equal weight model that used data only from sites where species were observed. The latter approach suffers from two drawbacks. First, outliers may significantly influence prediction, and second, the environmental data only comes from sites where species were observed so the explained variation in the results decreases, and hence the accuracy of prediction also decreases. Sperduto & Congalton, (1996) address this drawback by using the central part of the observed environmental values, and the second by developing the chi-square model that takes the data from sites presumed to be without modelled species. However, both approaches require more than the 1-6 input locations, which was the situation in this study, so were considered to be inappropriate. For the other five species modelled, the simple range overlapping, as used by Pfab & Witkowski, (1997) seems to have been effective.

The models presented in this paper are illustrative examples of the methodology, rather than the final predicted distribution maps of the modelled species. Accuracy assessments are still to be carried out, and as Cherrill et al., (1995) observe, the rarity of species is the factor which will probably have the greatest effect on the accuracy of model predictions.

However, rare species are very interesting to model because of necessity for their protection. Knowing real, or at least potential, distribution of the species is necessary for effective protection plan. Once developed, a predictive model could be used for discovery of previously unknown populations of modelled species like Sperduto & Congalton, (1996), or for prediction of changes in the species distribution caused by alteration in the environmental factors included in the model (e.g. Austin, 1998).

The Modelling of individual species is a more direct approach influenced mostly by physiological tolerance of species, ignoring competition between species. The advantage of extracting the most significant environmental variables by CCA analysis lies in fact that by analysing the floristic composition or co-occurrence of species, predictive models

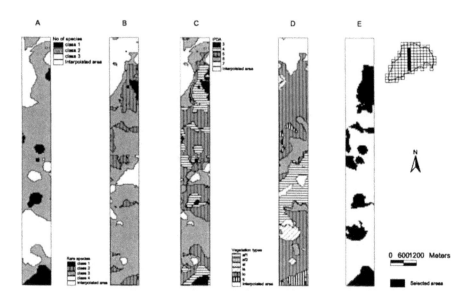

FIGURE 4. **A** - *predicted distribution of plant diversity based on number of plant species with class 1 indicating the greatest diversity,* **B** - *predicted simultaneous occurrence of six modelled species, with class 1 indicating highest number of simultaneous occurrence,* **C** - *result of spatial overlapp of maps from FIGURE 4A and 4B, with ipda index as a measure of plant diversity (sum of number of plant species index and simultaneous occurrence of rare species index),* **D** - *predicted distribution of six vegetation types, based on the results of cluster analysis (FIGURE 3),* **E** - *final selected areas that best represents plant diversity of survey transect.*

are likely to be more accurate in a spatial sense, since they take into account competitive interactions in nature, which all species have to deal with. Usage of abundance data would put even more weight on the role of competition in the distribution of modelled species.

Although a vegetation map for part of the surveyed area exists, the results of cluster analysis were preferred for determining the vegetation types based on the environmental profile of the clusters. This was for two reasons. First, a comparison of the real vegetation on surveyed plots with that attributed to them by the digitised vegetation map showed that the vegetation map was inaccurate in the vegetation that it represented. This is, in part, because of the difference in the minimum mapping unit and scale of the map compared to the detail with which the plots were surveyed. The second reason was an objective to cover the diversity of floristic composition within vegetation types. This was justified by forming two clusters in beech-fir forest (clusters C1 and C2 in Figure 3) that was treated uniformly at the association level by the criterion of Zürich - Montpellier phytosociological school. Diversity in the beech forest was covered by the presence of two associations (acid and moderate), while oak forest was homogenous, both confirmed by the cluster analysis. The modelled map of vegetation types (Figure 4D) showed some illogical distributions such as belts of beech forest between two types of beech-fir forests that do not occur in nature in this particular area, but this was only the result of gaps in the altitude range in data from beech-fir forests. Despite these belts, prediction of vegetation types based on floristic elements present was acceptable for most of the area.

The threshold value of 10 ha used was chosen according to Bergstedt (1994) who put the accent on the need for the existence of a buffer zone around such a minimum area, while Heinrich & Hergt (1994) state an area of 50-100 ha for the protection of vegetation. A polygon of acid beech forest, with at least one modelled rare or endangered species present, which satisfied the threshold value was added to areas classified as being of importance for plant diversity. This additional polygon, which satisfied both criteria, had an area of 55.23 ha. Alder forest (cluster D in Figure3) was not added to IPDA areas because of its small size and marginal position.

Areas that were previously defined as being rich in plant diversity and rare species were also tested with the same threshold value resulting in the exclusion of 16 polygons. Since the aim of this paper was to demonstrate the methodology and procedure that one might use for the selection of areas important for protection, final map (Figure 4E) should not be used as a specific protection plan for the surveyed area. The size of protected areas, their spatial patterns and shape are very important (Beardsley & Stoms, 1993), and GIS-based approaches enable the derivation and interpretation of such areas in a more robust and defendable manner. Protected areas should also be sufficiently distant to ensure the survival of species in case of catastrophe and yet close enough to ensure recolonization from preserved sites (Loomis & Echohawk, 1999). Additional criteria can be found in the approach proposed by Scott et al., (1993).

The spatial distribution of the surveyed plots (Figure 1B) probably resulted in overextending the predictive power of the model in some parts of the area enclosed by the interpolation. In future research, additional sources of spatial data will be considered for the provision of vegetation data, such as remote sensing. Based on such data, it will be possible to define the areas for field survey, balancing the variability in plant cover and the costs of fieldwork. Despite the high cost of fieldwork it is still essential for obtaining data about the ecology of particular species, especially rare ones. These data can then be used for developing predictive models to be applied on non-surveyed areas. Once developed, models could be used for anticipation of alterations in biological diversity caused by changes in the environmental factors.

Variables derived from the digital elevation model (Gottfried et al., 1998; Guisan et al., 1999) should also be included in the development of the models, and such variables could be used as indirect estimators for other environmental variables, such as estimators of the topographic exposure to wind representing level of exposure to air pollution source (Antonić & Legović, 1999).

6   Conclusions

The total number of plant species per 50 x 50 metres area was found to be positively correlated with topsoil acidity and concentrations of cobalt, chromium and yttrium, while the correlation with magnesium concentration was negative. Variables that gave the best descriptions of the variability in floristic composition among the plots were altitude, soil acidity, concentrations of nickel, cobalt and potassium. Cluster analysis of the floristic composition of the surveyed plots formed clusters in proportion with vegetation observed in the field, ensuring the correct definition of vegetation types on the surrounding area based on similar environmental conditions.

Plant diversity, expressed only through the number of species, did not prove a sufficient criterion for the identification of the areas that best represent plant diversity in a comprehensive manner. The floristic composition should also be included in this kind of analysis and in the formulation of decision making of plans for species protection.

The results presented here from the survey of Medvednica Nature Park provide a framework for the management of protected areas. Further modification of the method is required, as well as enlargement of the sample. Additional studies at different scales, and in different bioclimatic conditions, are needed to broaden the base with which to test the development of species distribution models. This will then provide a basis for the use of a combined approach of statistical modelling with GIS to produce an atlas of Croatian flora.

Acknowledgements

Dr Toni Nikolić from the Department of Botany, Faculty of Science, University of Zagreb initiated this research, ensured all the necessary software and hardware components, and was supportive all the time. I would like to thank Dr Vesna Lužar-Stiffler from the University Computing Centre in Zagreb, and Dr Oleg Antonić from the Institute "Ruđer Bošković" in Zagreb for advice about statistics. The Croatian Ministry of Science and Technology supported this research (Grant No. 119116, 1996/99). I also thank the anonymous reviewer whose comments significantly increased the clarity of this paper.

References

Antonić, O. & Legović, T. (1999). Estimating the direction of an unknown air pollution source using a digital elevation model and a sample of deposition. *Ecological Modelling*, 124(1), 85-95.

Austin, M. P. (1998). An ecological perspective on biodiversity investigations: examples from Australian eucalypt forests. *Annales of the Missouri Botanical Garden*, 85, 2-17.

Beardsley, K. & Stoms, D. (1993). Compiling a Digital Map of Areas Managed for Biodiversity in California. *Natural Areas Journal*, 13, 177-190.

Becker, B., Terrones, F. & Horchler, P. (1998). Weed communities in Andean cropping systems of northern Peru. *Angewandte Botanik*, 72, 113-130.

Bergstedt, J. (1994). Handbuch Angewandter Biotopschutz - Ökologische und rechtliche Grundlagen, Merkblätter und Arbeitshilfen für die Praxis. (pp. IV-3.1.1). Landsberg/Lech: ecomed Fachverlag.

Cherrill, A. J., McClean, C., Watson, P., Tucker, K., Rushton, S. P. & Sanderson, R. (1995). Predicting the distributions of plant species at the regional scale: a hierarchical matrix model. *Landscape Ecology*, 10(4), 197-207.

Gottfried, M., Pauli, H. & Grabherr, G. (1998). Prediction of vegetation patterns at the limits of plant life: A new view of the Alpine-nival ecotone. *Arctic and Alpine Research*, 30(3), 207-221.

Guisan, A., Weiss, S.B. & Weiss, A.D. (1999). GLM versus CCA spatial modeling of plant species distribution. *Plant Ecology*, 143, 107-122.

Heinrich, D. & Hergt, M. (1994). dtv - Atlas zur Ökologie. (pp. 228). München: Deutscher Taschenbuch Verlag.

Jelaska, S. D. & Nikolić, T. (2000). Geochemical control of the forest plant diversity on Mt. Medvednica, Croatia. *Periodicum Biologorum*, 102(2), (in press).

Kadmon, R. & Danin, A. (1997). Floristic variation in Israel: a GIS analysis. *Flora* 192, 341-345.

Kazda, M. (1995). Changes in alder fens following a decrease in the ground water table: results of a geographical information system application. *Journal of Applied Ecology,* 32. 100-110.

Loomis, J. & Echohawk, J. C. (1999). Using GIS to identify under-represented ecosystems in the National Wilderness Preservation System in the USA. *Environmental Conservation,* 26(1), 53-58.

McCune, B. & Allen, T. F. H. (1985). Will similar forests develop on similar sites? *Canadian Journal of Botany,* 63, 367-376.

Nikolić T (Ed.). (1994). Flora Croatica, Index florae Croaticae Pars 1. *Natura Croatica,* 3(Suppl. 2), 1-116

Nikolić T (Ed.). (1997). Flora Croatica, Index florae Croaticae Pars 2. *Natura Croatica,* 6(Suppl. 1), 1-232

Nikolić T (Ed.). (2000). Flora Croatica, Index florae Croaticae Pars 3. *Natura Croatica,* (in press)

Nikolić, T., Bukovec, D., Šopf, J. & Jelaska, S.D. (1998). Mapping of Croatian flora - possibilities and standards. (In Croatian, with summary, tables, figures and appendices in English) *Natura Croatica,* 7(Suppl 1), 1-62.

Pfab, M. F. & Witkowski, E. T. F. (1997). Use of Geographical Information Systems in the search for additional populations, or sites suitable for re-establishment, of the endangered Northern province endemic *Euphorbia clivicola. South African Journal of Botany,* 63(6), 351-355.

Podani, J. (1993). SYN-TAX-pc. Computer Programs for Multivariate Data Analysis in Ecology and Systematics. Version 5.0 User's Guide. Budapest: Scientia Publishing.

Preston, J., Engel, B., Lalor, G.C. & Vutchkov, M.K. (1996). The application of geographic infromation systems to geochemical studies in Jamaica. *Environmental Geochemistry and Health,* 18, 99-104.

Scholz, R.W., Nothbaum, N. & May, T.W. (1994). Fixed and hypothesis-guided soil sampling methods - Principles, strategies, and examples. In B. Markert (Ed.), *Environmental Sampling for Trace Analysis* (pp. 335-345). Weinheim: VCH Verlagsgesellschaft mbH.

Scott, J.M., Davis, F., Csuti, B., Noss, R., Butterfield, B., Groves, C., Anderson, H., Caicco, S., D'Erchia, F., Edwards, jr., T.C., Ulliman, J. & Wright, R.G. (1993). Gap analysis: a geographic approach to protection of biological diversity. *Wildlife Monographs,* 123, 1-41.

Sperduto, M. B. and Congalton, R. G. (1996). Predicting rare orchid (Small Whorled Pogonia) habitat using GIS. *Photogrametric Engineering & Remote Sensing,* 62(11), 1269-1279.

StatSoft, Inc. (1995). STATISTICA for Windows [Computer program manual]. Tulsa, OK: StatSoft, Inc., 2325 East 13th Street, Tulsa, OK 74104, (918) 583-4149, fax: (918) 583-4376.

Ter Braak, C. J. F. (1988). CANOCO - a FORTRAN program for canonical community ordination by [partial] [detrended] [canonical] correspondence analysis, principal components analysis and redundancy analysis (version 2.1). Technical report: LWA-88-02. Wageningen: Agricultural Mathematics Group.

Ter Braak, C.J.F. (1990). Update notes: CANOCO version 3.10. Wageningen: Agricultural Mathematics Group.

12

Interpolation In A Heterogenous World

David M. Kidd

1.   Introduction

The world is a heterogeneous place across which geographical processes produce spatial patterns. Causal relationships between processes and landscape features result in relationships between pattern and landscape. Spatial patterns are often modelled by interpolation of a surface from sample data. However, despite recognition of the relationships that exist between process, pattern and landscape few interpolation techniques allow the incorporation of landscape features in the surface modelling process. This paper briefly reviews the ability of several commonly used interpolation functions to incorporate landscape features. Particular attention is given to functions that allow the inclusion of barriers in surface generation. Limitations in how barriers are implemented lead to the development of a new 'hybrid' interpolation technique network surface interpolation (NetSURF). NetSURF is compared to existing functions using both a test data set and a real world example, the generation of organism trait surfaces.

2.   Interpolation Techniques And Landscape Heterogeneity

Relationships between process and landscape can be broadly classified into *covariation*, *transition* and *barrier* relationships. *Covariation* relationships are where landscape and process interact (positively or negatively) to reinforce a pattern, e.g. plant species abundance may be determined by the availability of a single nutrient. *Transition* relationships are gradual or abrupt changes in landscape form that alter the relationship between process and pattern, e.g. a change in slope may alter community composition by altering soil drainage. *Barriers* are where landscape features alter process flows. Barriers may be absolute (prevent all flow) or partial (restrict flow), e.g. rivers, seas and mountains can be barriers to organism dispersal and hence gene flow. The term 'barrier' is perhaps not a good choice of nomen for this class of relationship as 'barriers' could enhance, instead of diminish flow, e.g. organisms may move through many different habitat types but preferentially use river corridors.

Simple bivariate transitions and covariation relationships can be regarded as two perspectives on the same situation – areas of covariation are separated by transitions. When this is the situation the choice of model, covariation or transition, will probably depend on survey design, the format of available data sets and the functionality of available software. In more complex circumstances where a transition is the manifestation of several interacting processes it is more difficult, and may be impossible, to represent the transition with a multivariate covariation model.

Technique										
Software		Triangular Irregular Network (TIN)	Inverse Distance Weight (IDW)	Spline	Rectangular Interpolation	Trend Surface	Kriging	Bilinear	Nearest Neighbour	Radial Basis Function	Shepard's Method
	ArcInfo 7.1.2	B, K	B	Y	.	Y	B	.	.	.	.
	ArcView (Spatial Analyst 1.1, 3D Analyst)	K	B	Y	.	.	.	.	.	.	.
	IDL 5.2	Y	.	Y	.	.	C	Y	.	.	Y
	Idrisi for Windows 1.01.006	.	Y	.	.	Y	.	.	.	.	.
	MapInfo (Vertical Mapper 1.5)	Y	Y	.	Y	.	.	.	.	.	.
	Infrasoft MX 2.3	Y	.	.	.	.	.	.	.	.	.
	Surfer 6.04	Y	Y	Y	.	Y	Y	.	Y	Y	Y

Table key:
., technique not supported.
Y, technique supported but no landscape features supported.
C, cokriging (covariation) supported.
K, breaklines (transition) supported.
B, barriers (barriers) supported.

Table 1. *Incorporation of landscape features within interpolation functions available within the Department of Geography, University of Portsmouth, June 1999.*

The typology of covariation, transition and barriers reflects how landscape heterogeneity is incorporated in surface algorithms, table 1. This typology should in no way regarded as a definitive classification. Table 1 shows that very few interpolation functions allow the incorporation of any type of landscape feature within the modelling process. Only Environmental Systems Research Institute's (ESRI) Arc products allow the inclusion of barriers and breaklines. ArcInfo and ArcView inverse distance weighting (IDW) both support barriers and triangular irregular network (TIN) breaklines, whereas, only ArcInfo TIN supports barriers. Of the software investigated only the Interactive Data Language (IDL, RSI 1999) supports cokriging, however, this is not in any way representative of the availability of this technique as no specialist geostatistical software were examined. The remainder of this chapter examines in detail the implementation of barriers in the Arc products then describes a new approach to modelling barriers – the 'hybrid' NetSURF technique.

3.   Barriers In Arcinfo And Arcview

Arc products only support absolute barriers. ArcInfo TIN and ArcView and ArcInfo IDW functions were tested using a simple data set that mimics a dividing mountain range and pass. The data set consists of low and high values on either side of a barrier dissected by a sigmoidal passage within which there are no samples (figure 1). Also depicted in figure 1 is the position of the long-sections shown in figure 4.

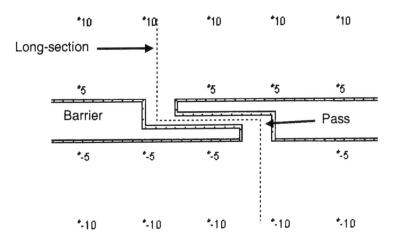

Figure 1. *Test data set.*

IDW surfaces were generated from the test data set using the three nearest sample locations weighted inversely by distance. These parameters were chosen to make as direct a comparison as possible with the TIN model which uses the z values of the three triangle apexes to interpolate z values at locations within the TIN facet. Both ArcInfo and ArcView IDW with barriers produce the same output from the test data set (figure 2).

Arc IDW uses a visibility rule to determine which sample points are used to calculate grid cell values (figure 3). With the test data this results in NO DATA values being assigned to the centre of the passage from where no sample locations are visible. Other areas have values biased towards visible rather than the nearest sample points. This is the reason for the +10 and –10 areas found to either side of the NO DATA area; only sample points opposite the mouth of the barrier passage are visible from these areas. These artefacts are clearly visible in the long-section where from each end of the section z values gradually converge towards zero before rising to 10 then stopping (figure 4). Contours generated from the surface show triangular patterns in the region outside the passage due to only the nearest three points being used. In addition, a confusing pattern of contours is seen in the passage mouth where small changes in position result in changes in the set of points visible from that location.

An ArcInfo linear TIN surface was also generated from the test data set (figure 5). For clarity the TIN surface interior to the barrier has been removed with a hard erase polygon of the same shape as the barrier. It is difficult to exactly reconstruct how the TIN with barriers surface is created from the ArcInfo documentation (ESRI 1997). The documentation does however provide the following three rules; (i) 'z values along each side of a barrier feature are automatically interpolated .... from the z values of all other input features that are on the same side of the barrier', (ii) the '...model surrounds each barrier arc with a sliver-shaped region of transition; locations inside this region have the NODATA value', and (iii) barriers cannot have z values at nodes, only NODATA values. How these rules result in the test data TIN surface observed is not obvious.

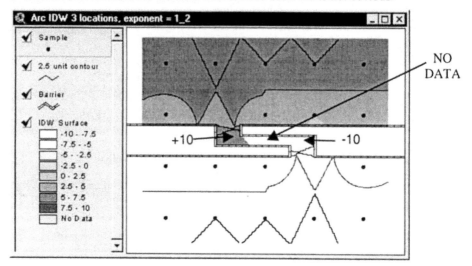

Figure 2. *Arc IDW Surface*

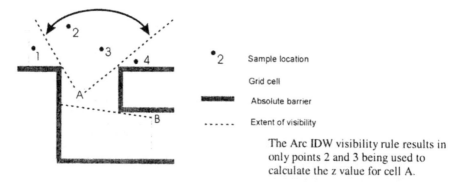

2   Sample location

             Grid cell

━━━   Absolute barrier

- - - - -   Extent of visibility

The Arc IDW visibility rule results in only points 2 and 3 being used to calculate the z value for cell A.

In contrast, cell B is assigned the no data values as no sample locations are visible from it.

Figure 3. *The Arc IDW visibility rule.*

Comparison of figures 2 and 5 show that in contrast to Arc IDW, ArcInfo TIN interpolates values along the entire length of the passage. However, within the passage the surface slope is so shallow as to be effectively flat. Z values along the lower central wall of the passage (marked 'A' in figure 4) only range between -0.0845×10^{-5} and $+1.63 \times 10^{-5}$, just over one ten-millionth of the variation in input sample location z values. The TIN with barriers exhibits a number of other undesirable artefacts. Sudden changes in TIN node values occur at barrier corners, three examples of which can be seen in the central and right-hand details of figure 6 where there are abrupt changes in z values between 'wall' and 'corner' sections of the TIN where it is offset from the barrier.

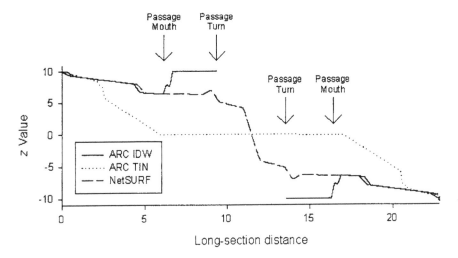

Figure 4. *Long-sections through barrier passage.*

Second, the 'inner-arms' of the passage, marked 'B' in figures 5 and 6, have much larger z values than the other passage walls. The values of the wall shown in figure 6 range from 0.0296 to 0.2072, much lower than the sample location values but much larger than other TIN node z values in the passage that are in the $\pm 0.1 \times 10^{-5}$ range. In addition, node values along the inner-arms increase the in opposite direction to that expected. The reasons for these discrepancies remain elusive without additional information on the mechanics of the algorithm.

A further undesirable artefact of the TIN model are differences in the topology of the triangulation where it is offset from seemingly identical barrier corners. This can be seen in figure 6 where the left detail shows a different TIN configuration to the other details despite all of them being the consequence of a simple right angle in the barrier. Contours produced from the TIN do not show the triangular form or 'confusion' in the passage mouth observed with IDW. Instead we see 'faults' in front of the passage mouth where z values change abruptly.

The use of barriers in interpolation clearly has great potential as a means of incorporating landscape structure and spatial processes in surface interpolation. However, the barrier techniques investigated only support absolute barriers and produce outputs with a number of artefacts that restrict their application.

4.   A Routing Problem

Barriers interrupt or modify process flows across the landscape; partial barriers restrict flow while absolute barriers require circumnavigation. Estimation (interpolation) of z values at unsampled locations requires the 'nearest' surrounding locations to be found taking into account the barriers; this is essentially a routing problem. There are two basic techniques to routing problems which rely on different spatial models; a route can be *threaded* across a friction surface or *traced* along a network.

Threading seems the most natural technique for phenomena that are relatively unrestricted in flow direction (not limited to a network of some kind, e.g. road or river). There are several ways to thread a path across a surface. Bowser (1996) used standard cost path analysis to assess the effect of environmental isolation on genetic diversity between Black-tailed prairie dog colonies; in contrast, Scippers et al. (1996) used replicate random walks to assess isolation between European badger sets. Both these methods were successful for calculating small numbers of pair-wise isolation distances (7 prairie dog colonies and 20 badger sets respectively). However for surface interpolation cost paths between all grid cells and all sample sites would need to be calculated, doing this would be extremely computationally intensive and time consuming even for small grids.

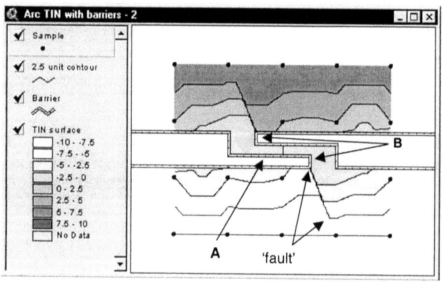

Figure 5. *Arc TIN Surface*

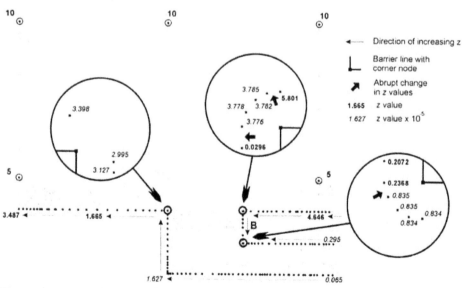

Figure 6. *TIN with barrier node z values (only top-left of test data shown). Details show TIN offset at barrier corners.*

In addition to the computational practicalities of path threading there are two issues of path quality that must be considered. The first is that the grid spatial model limits path orientation to the eight king's move directions which often results in paths longer than if a vector spatial model were employed (figure 7). Secondly, Bowser (1996) identified non-reciprocity in least cost paths (i.e. the path from X to Y may not be the same path as Y to X) leading to uncertainty that the shortest path has indeed been identified.

The problems with path threading stimulated the examination of network tracing as a possible alternative method for identifying minimum cost routes. Use of a network spatial model imposes constraints on the set of possible paths; however, we can design our network to minimise this limitation. A distinct advantage of network tracing over cost path treading is that it is considerably more computationally efficient at calculating multiple least cost routes.

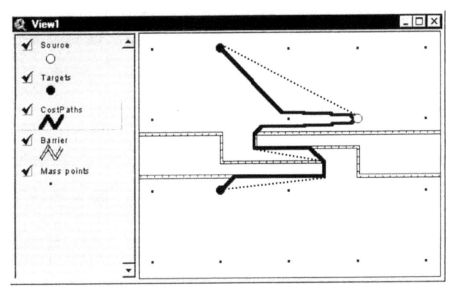

Figure 7. *Problems with cost paths.*
Two paths generated with ArcInfo COSTDISTANCE and COSTPATH (all cells outside the barrier have a cost of 1 unit, barrier is absolute). Both paths are clearly longer than the source-target minimum distance with a vector spatial model (dashed lines)

5.   The NetSURF Technique

Network Surfacing (NetSURF) is a novel 'hybrid' interpolation technique that has been developed to generate surfaces incorporating both partial and absolute barriers. The method is described as 'hybrid' as it uses a variety of GIS functions not normally used together. NetSURF is applied to both the test data set and a real-world example - interpolating trait surfaces for the Saddle-backed Bushcricket in Western Europe. NetSURF is perhaps best understood by breaking the technique into four stages, network design and construction, network distance calculation, node interpolation and surface construction.

The first stage is to design and construct the network that is used to calculate z values at the network nodes on the basis of network distance to the nearest x sample locations.

The network must be designed to serve three functions, (i) calculate minimum cost distances, (ii) be a model of landscape variation, and (iii) act as a sparse 'space-filling' framework. The space-filling function increases network connectivity (making distance calculations more accurate) and provides a framework of spatially distributed mass points from which a continuous surface can be created.

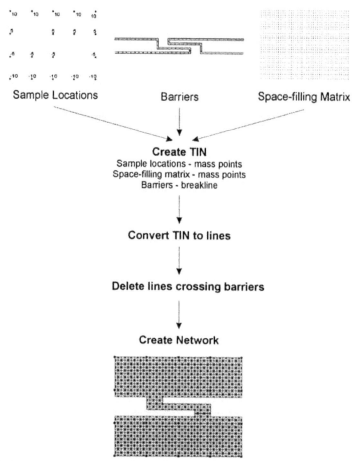

Figure 8. *Creating the test data NetSURF network.*

The network for the test data set was created from three data sets, sample locations, barriers and a matrix of space-filling points (figure 8). A TIN was generated from these data sets to create network links between the point data sets and line nodes. The TIN was then converted to a line coverage and links crossing the absolute barriers deleted. If the barrier had been partial instead of absolute a friction cost would have been assigned to the links crossing the barrier based on barrier strength. An alternative to deleting lines crossing absolute barriers would be to assign them infinite cost (in Arc the NODATA value). The final step of the first stage is to create the network route system from the line coverage with the ArcPlot NETCOVER command.

The second stage is to calculate the shortest network distances between all network nodes and sample locations with ArcPlot NODEDISTANCE. The file of network distances is then exported to a custom program written in the IDL (RSI 1999) language. This program completes the third stage, interpolation of network node z values. Z values are calculated for all network nodes using a user specified distance weight function on the x nearest sampled locations. In the test example described here a simple inverse linear weighting was applied on the nearest three sample locations; the same parameters as previously used with Arc IDW. The IDL program outputs a text file of interpolated node z values. The final stage is to import this file back into ArcInfo and create a TIN surface using the node z values as mass points again erasing the area enclosed by the barriers (figure 9). The entire NetSURF procedure could have been written as an ArcInfo AML macro instead using IDL. IDL was only used pragmatically to speed up development.

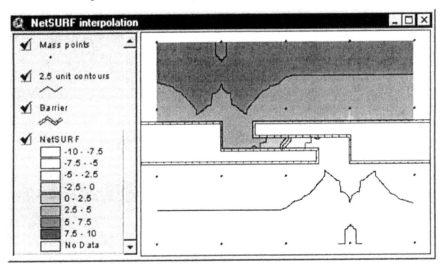

Figure 9. *NetSURF surface from test data set, 3 nearest nodes linear with distance*

NetSURF appears to produce a more satisfying output than either Arc IDW or Arc TIN. In particular in the contour lines do not exhibit either the triangular patterns seen in the IDW surface (figure 2) or the fault lines of the TIN surface (figure 5). Z values are interpolated smoothly along the entire length of the pass (figures 4 and 9). The steep slope seen in the centre of the passage is an artefact of the sample location distribution in the test data set. The regular pattern and the low number of sample locations (three) used in the interpolation result in rapid change in the 'nearest' sample locations over short distances in this area. A more irregular distribution of samples or a greater number of samples in the interpolation algorithm would have diminished this.

6.   A Real World Application Of NetSURF

Trait patterns within species (intraspecific variation) are the product of several evolutionary spatio-temporal processes including, developmental plasticity and natural selection of individuals, genetic drift within populations, gene flow between populations, and historical population constraints. Landscape heterogeneity directly influences these processes, for example, habitat quality and area determine local population size, whereas, habitat distribution determines the isolation of populations from one another. Population

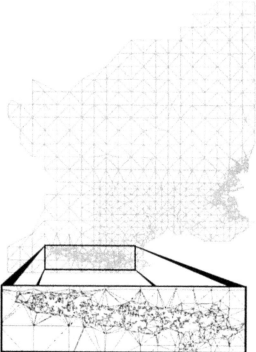

Figure 10. *User defined network for Bushcricket NetSURF interpolation with detail showing mountain passes in the Pyrenees.*

size, in turn, influences the importance of drift within populations while as isolation between populations increases so does the potential for local differentiation. Providing the relationships between landscape, process and pattern are sufficiently understood the inclusion of such factors in the interpolation of trait surfaces from samples has great potential to improve modelled surfaces. Gene flow between populations is the consequence of dispersal of individuals, gametes (e.g. pollen) and propagules (e.g. seeds) whose passage is slowed by partial barriers and halted by absolute barriers in the landscape.

The Saddle-backed Bushcricket (*Ephippiger ephippiger*) has a disputed taxonomy and is highly variable for a variety of traits in Southwest Europe from Northern Spain through France and the Rhône valley to the Western Alps (see Ritchie et al 1997 for a brief review). Within this region the Mediterranean Sea and mountain regions in the Western Alps and Pyrenees above approximately 2050m are believed to be absolute barriers to dispersal. Kidd and Ritchie (2000) generated IDW surfaces for a number of traits with the Idrisi GIS (Eastman 1992) from which they inferred possible evolutionary scenarios which could have produced the observed patterns. Idrisi IDW surfaces cannot incorporate barriers so NetSURF surfaces were produced to evaluate if the inclusion of barriers would significantly effect the observed patterns.

The bushcricket network was constructed in the same way as the network for the test data set. Three data sets were input, sample locations, barriers (2050m contour derived from a 30 arc-second DEM, USGS 1996) and a set of space-filling points (figure 10). To reduce network size and complexity the density of the space filling points was varied with the density of sample locations and some network links to the east of sampled locations in the Alps were deleted. After editing the network consisted of 3257 nodes and 7434 links. With the 124 sample locations this results in 403,868 sample location–node distance pairs.

NetSURF surfaces were generated for thirty-one traits using the six nearest sample locations inversely weighted by distance. Here only the NetSURF surface for average length of the pronotum is compared to a standard Arc IDW surface (figure 11). The pronotum is the dorsal surface of the first segment of the thorax which bears the anterior (first pair) of legs. In bushcrickets the pronotum is often modified into a species-specific shape; hence the common name of *E. ephippiger* is the Saddle-backed Bushcricket. Inclusion of the Pyrenean mountain barrier has had a small effect on interpolated *z* values in the Eastern Pyrenees with average pronotum length comparatively higher to the south of the watershed in the NetSURF compared to the standard IDW surface. The Mediterranean Sea seems to have had little effect. Other traits modelled with NetSURF also showed similar pattern changes with increased differentiation between France and Iberia, especially male calling song. In general the NetSURF patterns better match a post-glacial secondary contact evolutionary scenario than the simple IDW surfaces (Kidd and Ritchie, 2001). An additional benefit of the technique for studies investigating evolutionary scenarios is that the matrix of 'as-the-cricket-crawls' inter-sample site distances can be easily extracted from the sample location-node distance file. This matrix then be can used in Mantel Tests to assess the significance of isolation-by-distance (drift) over other evolutionary hypothesis (Ritchie et al, 2001).

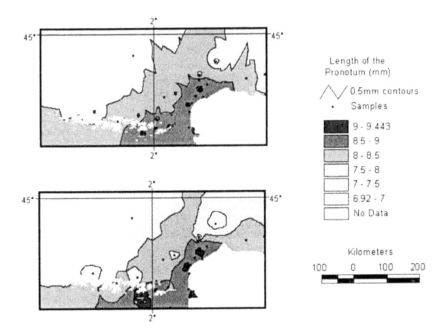

Figure 11. *Comparison of ArcIDW and NetSURF surfaces for average length of the pronotum in Southeast France and Northern Spain. Both surfaces use the 6 nearest sampled locations inversely weighted by distance. ArcView IDW, (b) NetSURF.*

7. Summary

Despite the recognition that landscape can significantly effect spatial processes and hence patterns, a limited review of interpolation functions revealed that few allow the inclusion of covariation relationships, transitions or barriers in the surface modelling process. Trials of functions supporting barriers (ArcInfo TIN and IDW, ArcView IDW) reveal limitations in their implementation; only absolute barriers are supported, local NO DATA or flat zones are produced and the algorithms produce areas where z values change abruptly resulting in poor contour definition. The net effect of these problems is to make these functions only suitable for representing simple situations.

These problems stimulated the development of a 'hybrid' interpolation technique, network surface interpolation (NetSURF), which can incorporate both absolute and partial barriers. NetSURF uses a user-defined network to find least cost routes between sample locations and a set of space-filling network nodes, a user-defined formula can then be used to interpolate z values at the network nodes using the network distances as weights. A continuous surface is generated from the interpolated node z values. In contrast, to the Arc functions, NetSURF processing time is low and does not greatly increase with barrier complexity. Most importantly the technique does not produce the obvious artefacts, including abrupt changes and plateaux, exhibited by the Arc functions. The influence of network design on the generated surfaces remains to be investigated, in particular, visibility graphs (O'Rourke 1994) could be used to ensure minimum paths between sample locations are incorporated in the network.

NetSURF has been used to generate trait surfaces for the Saddle-backed bushcricket the results of which have helped improve our understanding of the relative importance of the evolutionary processes acting on this species. It is now proposed to develop an Arc AML macro freeing the technique from the IDL programming environment. NetSURF is a small step towards employing our understanding of environmental process to embed significant landscape features within spatial interpolation to produce more realistic surface models that recognise that the world is a heterogeneous place.

Acknowledgements

I wish to thank Shane Murnion for suggesting network tracing as a possible way to proceed and Michael Ritchie and Matthijs Duijm for the use of the bushcricket data. Also David Livingstone and Richard Healey for their support and critical reading of this paper and the anonymous referee for their useful comments.

ArcInfo and ArcView are registered trademarks of the Environmental Systems Research Institute, Inc. 380 New York Street, Redlands, CA 92373-8100, USA. IDL is a registered trademark of Research Systems Inc. 4990 Pearl East Circle, Boulder Colorado 80301, USA.

References

Bowser G, 1996. Integrating ecological tools with remotely sensed data: modeling animal dispersal on complex landscapes. Third International Conference/Workshop on Integrating GIS and Environmental Modeling. National Center for Geographic Information and Analysis, January 21-25, 1996 Santa Fe, New Mexico, USA
Eastman JR, 1992. Idrisi 4.0. Clark University, Massachusetts, USA.

Environmental Systems Research Institute, Inc, 1997. ARC Version 7.1.2. Environmental Systems Research Institute Inc., 380 New York Street, Redlands, CA 92373-8100, USA.

Kidd D and Ritchie M, 2000. Inferring the patterns and causes of geographic variation in *Ephippiger ephippiger* (Orthoptera, Tettigonioidea) using Geographical Information Systems. Biological Journal of the Linnean Society 71, p 269-295.

Kidd D and Ritchie M, 2001. A geographical information science (GIS) approach to exploring variation in the bushcricket *Ephippiger ephippiger*. GIS and Remote Sensing Applications in Ecology and Biogeography. Dordrecht: Kluwer.

O'Rouke J, 1994. Computational Geometry in C. Cambridge University Press, Cambridge, UK. p 271-273.

Ritchie MG, Racey SN, Gleason JM and Wolff K, 1997. Variability of the bushcricket *Ephippiger ephippiger*: RAPD and song races. Heredity 79, p 286-294.

Ritchie MG, Kidd D and Gleason JM, 2001. mtDNA variation and GIS analysis confirm a secondary origin of geographic variation in the bushcricket *Ephippiger ephippiger* (Orthoptera: Tettigonioidea), and resurrect two subspecies. Molecular Ecology 10. Research Systems Inc. Interactive Data Language. Version 5.2. Research Systems Inc. 1999.

Scippers P, Verboom J, Knappen JP and van Apeldoorn RC. 1996. Dispersal and habitat connectivity in complex heterogenous landscapes: an analysis with a GIS-based random walk model. Ecography 19, p 97-106.

USGS 1996. GTOPO30, United States Geological Survey. http://edcdaac.usgs.gov/gtopo30/gtopo30.html.

13

Spatial Modelling Of The Visibility Of Land Use

David Miller

The integrated management of land use requires change in land cover to be managed with regard to its ecological, economic and landscape functionality. An inventory of resources that includes the contribution made by land cover to the visual landscape can be used in the monitoring of the implications of change in land cover on the visual landscape capital of an area. A method is presented for assessing the visual resources of the landscape, as applied to the western part of the prospective Cairngorm National Park in Scotland, including an illustration of the changes in diversity of view content and the change in and the area from which woodland is visible along a chosen route. The results show a reduction in the overall contribution of scattered, natural woodland to the visual landscape, and an increase in the visual contribution of plantation woodland. The limitations of the method are discussed, and where there is potential for providing a spatial context for expressing the results of landscape preference modelling.

1   Introduction

Existing land use has evolved for cultural, ecological and economic reasons (Wascher *et al.*, 1999), and the modelling and representation of the resultant landscape, in rural areas, is being carried out in response to identifiable needs for a greater understanding of the processes of change, with the objective of managing such change. The visual landscape provides one important element to the wider issue of the sustainability of the evolution of the landscape (Bell, 1999; Mladenoff and Baker, 1999; Daniel, 1999).

The development of techniques to assess the visibility of land use types can enable measurement and inputs to the evaluation of the potential impact of changes in land use upon visitor experience of the landscape. Comparisons of the data between different dates can also be analysed in a more targeted fashion, such as with respect to visitor view points, or particular focal points, such as bridges or footpaths. The application of such techniques include those for which a measure of geographic space is required including environmental impact assessments or landscape evaluation and classification (Kliskey *et al.*, 1994; Barbyn, 1996). The approach adopted is often one with which human beings may relate most directly, *i.e.* visualization or intervisibility calculations (Brown and Daniel, 1987; Coeterier, 1994), but these are usually based upon single or multi-directional views.

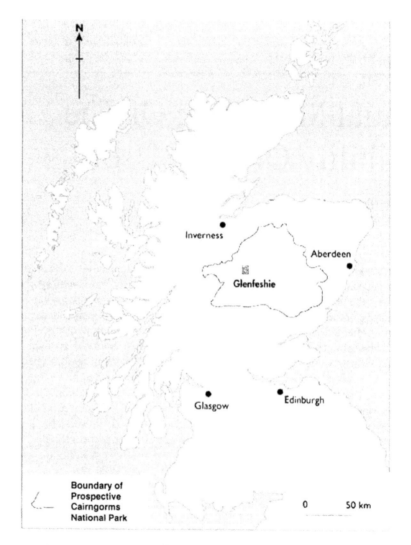

Figure 1. *Boundary of prospective Cairngorms National Park and location of Glenfeshie study area*

This paper presents a method for measuring and analysing the visibility of a land cover type (defined as meaning the land area from which a land cover type is visible) for different dates and assessing the change in levels of visibility of land cover, or use. The study is within the Cairngorms area of Scotland, soon to be designated as a National Park, for which the interactions between different land uses will require to be understood, as part of the strategic and tactical management of the area. The outputs from the analyses are principally estimates of the composition of the visual resource across the area, and how they can change over time, which can be used in the quantification of the visual impact of land use changes, but also a method for use by the putative National Park authority with commercially available Geographic Information Systems (GIS).

2 Background

2.1 Land Management

Management of the prospective Cairngorms National Park (Figure 1) will be split into four sectors: protection and conservation of the high land; protection and regeneration of native woodland, management of deer and maintenance of managed heather moorland; social and economic well-being of local communities (including recreation and access); and nature conservation and landscape (Cairngorms Partnership, 1999). Among the most significant of the natural heritage features contributing to both the habitat and the visual landscape are the extensive areas of ancient semi-natural woodland (Peterken *et al.,* 1995), and their spatial juxtaposition with the heather moorland.

Within the vision for the landscape strategy for the wider countryside is a 'living landscape' (i.e. the recognition that it must support economic activity, as well as respect the significance of the conservation of natural and cultural heritage features), with a dynamic mixture of land uses including forestry, heather moorland, agriculture and wetlands, interspersed with country houses and villages. Throughout the area, the land management should be sensitive to the design of man-made features, and minimize those that are unsightly or diminish local identity, including the contribution of spatial distribution and juxtaposition of land cover types and uses.

The Cairngorms exhibit several gross, man-induced, changes in land cover such as afforestation and the introduction of new hill tracks. Of particular interest in this paper, is the contribution made to the visual landscape of the extent and distribution of plantation woodland, and the more open, scattered woodlands (Gourlay and Slee, 1998; Wherrett, 1998), and the change in visibility of land use types from tracks and paths.

2.2 Landscape Visibility

The derivation of the viewshed, from individual points, has provided the spatial expression of visibility within which the measurement and analyses of landscape features have been made for many years. Much of the more recent research into the viewshed focuses upon algorithms and the assessment of accuracy of the intervisibility of features (DeFloriana and Magillo, 1994; Fisher, 1994; 1996). However, the spatial analysis of the structure and patterns of landscape visibility is less well studied, and the spatial modelling of the distribution of the visible area of features, based upon samples or census approaches to feature visibility, although used in architecture and urban environments (Benedikt, 1979) is little applied in the study of rural landscapes, such as those in the Cairngorms.

Benedikt (1979) describes the role of 'space' in architectural design, which closely relates to work by Tandy (1971), both of which seek to explain the importance of understanding what the perception of space contributes to an individual in terms of their 'comfort' and preference with respect to living areas. By measuring the length of the lines of sight from the observer to each corner of a room and accounting for obstructions to a line of sight, the proportion of a room which was both visible and thus 'open' could be calculated, described as an "isovist". Steinitz (1990) applies a similar approach to the calculation of the extent and content of views from a vehicle travelling along a set route, producing an output which is raster dataset that could be used in conjunction with other such datasets.

The means for identifying and describing the presence of cultural or land cover features has been dealt with by several authors using site visits, photographs, graphical and digital maps (Coeterier, 1990), and subjective scoring systems are often employed for relating the extent of the presence of such features in the landscape.

The handling of geographic data in a digital form has provided different ways of quantifying scene content and the use of visibility functions within GIS provides the basic means of estimating the extent and location of land which is visible to an observer or from where a feature may be visible (DeFloriani and Magillo, 1994; Fisher, 1996). However, assessing the extent to which features are visible in the landscape, or indices of landscape visibility have also been reported by authors including, Franklin and Ray (1994), Miller and Law (1997) and Kidner et al. (1997).

DeFloriani and Magillo (1999) describe the use of TINs data structures for the derivation of surfaces of 'continuous visibility' (based upon the area of the view-shed related to a facet of a TIN), with particular attention to the nature of the algorithms used in its derivation. They note that the time taken to undertake such analyses is sensitive to dimensions and detail of the input data, and that using regular sized grids can be computationally expensive, suggesting that parallel processing of the data may be an appropriate means of improving the efficiency of the analysis (Teng et al., 1993).

The method, presented in this paper, for calculating the visibility of land cover types across large geographic areas uses regular sized grids to provide one means of comparing visibility between sites, which extends work on the census of visible land to include assessments of the visibility of land cover features, and their changes through time (Miller et al., 1994). The principal objectives are to enable the monitoring of changes in landscape, and the contributions made by different land cover types, by the prospective National Park authority (a successor to the Cairngorm's Partnership) and stakeholders including Scottish Natural Heritage and the Scottish Executive.

2.3 Study Area and Data

The study area is Glenfeshie, in the western part of the Cairngorm Mountains (Figure 1), which is centred on a valley that drains to the north into the River Spey. The area includes a part of the mountain plateau to the east and south and a range of land cover types of forestry, heather moorland and low intensity agricultural activities. It also provides a popular location for hill walking and recreation.

2.4 Land cover

Data on the current and historical land cover for the area were manually interpreted from aerial photography for 1946 and 1988 (MLURI, 1991; Hester et al., 1996). The 126 land cover features identified in the classification scheme were used as a basis for the interpretation, from which a total of 1327 classes were recorded. However, for this study the land cover data were recoded to a simplified scheme of 15 classes, in which the distinction between different elements of the land cover that may be most significant to the observer were recorded (e.g. density or type of woodland cover). Table 1 summarizes the land cover types used, and the extent of each within the study area of Glenfeshie.

2.5 Digital Elevation Models and orthophotographs

Two Digital Elevation Models (DEMs) were produced by digital photogrammetric techniques (horizontal resolution of 1 m x 1 m), from vertical aerial photography for

Land Cover Class	1946 Area (ha)	% of Area	Area of Visibility (ha)	Area (ha)	1988 % of Area	Area of Visibility (ha)
Agriculture	9.1	0.3	733.4	8.0	0.2	719.2
Water	1.8	0.0	18.5	1.7	0.0	18.6
Grassland	41.4	1.2	1 154.9	272.3	7.7	2 437.8
Built-up	2.0	0.1	295.2	0.0	0.0	0.0
Woodland	130.5	3.7	2 639.5	518.1	14.7	3 328.7
Heather	1 441.4	40.9	5 182.5	1 064.5	30.2	5 182.0
Peatland/Montane	485.7	13.8	4 519.5	498.8	14.1	4 557.8
Heather Mosaic (no trees)	23.4	0.7	1 093.1	25.7	0.7	1 308.4
Heather Mosaic (with scattered trees)	861.9	24.4	4 680.4	691.4	19.6	4 766.3
Peatland/Montane Mosaic	219.4	6.2	3 848.3	231.4	6.6	3 868.8
Grassland Mosaic (no trees)	157.7	4.5	2 083.2	58.0	1.6	1 207.7
Woodland Mosaic (woodland dominant)	75.2	2.1	1 431.9	156.0	4.4	1 961.9
Other/No data	76.3	2.2		0.0	0.0	
Total	**3 525.9**	**100.0**		**3 525.9**	**100**	

Table 1. *Extent of land cover classes and the area of visibility (i.e. the area of land from which the land cover type is visible) for Glenfeshie for 1946 and 1988. (Land cover class mosaics are coded according to the dominant land cover class. 'Other/No data' includes bare rock and missing aerial photography or areas obscured by cloud cover).*

1946 and 1998 (Miller, 1999). The use of a digital photogrammetric approach produces a raster dataset which also records the heights of surface features such as woodland canopies and certain geomorphological features (such as moraines) which will impact upon the calculation of the visibility of adjacent land, and not be represented within the national 1:10 000 DEM produced by the Ordnance Survey (Ordnance Survey, 1997).

The area of the study for which the visibility of land cover types was calculated was extracted from the centre of the total area of coverage of the DEMs and land, leaving a distance of 10 km to the nearest edge of the data. This ensured that the outputs from the visibility calculations were derived from a consistent land area. Therefore, the actual study area is approximately 5% of the total extent of the datasets.

3   Methodology

3.1 Derivation of land cover visibility datasets

The two main objectives of the analyses are to produce an inventory of the visibility of each land cover type, and an assessment of the change in the area from which different land cover types may be visible, between 1946 and 1988. The method uses the DEM and land cover datasets as inputs to the calculation of how much of each land cover type is visible from a regular sample of points from across the area (Miller and Law, 1997). The basis of the calculation is a count of the number of cells that are visible from all other cells within a preset radius of view, and by incorporating data on land cover, the outputs is an estimate of which land cover types are visible from each cell.

The assumptions made include: no representation of individual trees or buildings; no account taken of the ability of the observer to distinguish between different land cover types; no account is taken of the significance of factors that affect the view, such as atmospheric attenuation of colours and viewing angle; no account is taken of the effect of the density of tree cover on the observer to view out from a location.

3.2 Data Processing

The derivation of the area of land from which each land cover type was visible was undertaken in two steps.

1. The land cover data were converted from a vector into a raster format, in the 15 class scheme chosen, with a cell size equivalent to that of the DEM (1 m x 1 m).
2. The land cover dataset and the DEM were used as inputs to a FORTRAN routine, which calculates the intervisibility between each cell and every other cell in the land cover dataset. These intervisibility calculations used a maximum radius of view of 10 km radius, with an observer height of 1.8 m.

The management of the data processing concerned with the balance between the number of repeated intervisibility calculations, the memory required, the disk space, the time taken for each segment and the requirements of other users of the computer system. To improve the through-put of the data processing, the raster land cover dataset was recoded into four, with 4, 4, 4 and 3 land cover classes in each of the new datasets respectively, ensuring that there was only one land cover class represented in each cell, to avoid any possibility of counting the visibility of a cell more than once. The intervisibility calculations were then run for each dataset independently, using Sun Ultra 60 and Sun Ultra 40 workstations, each with 2 GBytes of memory and 2 x 360 MHz processors, and with one set of analyses running on each processor, and the DEM and land cover data loaded on the local hard disk. This approach to distributing the processing enabled the time required for completion to be reduced from approximately 12 days to 3.5 days of elapsed time and approximately 13 days of CPU time. A direct comparison of the time taken for the two approaches was not reliable because of the different demands made upon the hardware by other users. However, the computational costs were deemed acceptable for three reasons.

1. Of greatest significance was the requirement to derive a dataset that could be used for the purposes of the application, namely, an analyses of the changes in the visibility of different land cover types, and the lapsed time was not an inhibition to the successful creation of the required datasets.
2. The potential users of the data, and routines (the prospective National Park authority and the relevant stakeholders) use ERDAS IMAGINE (ERDAS, 1999) and ArcInfo (ESRI, 2000) as standard software packages, and required access to methods that would employ cell-based processing.

3. The basic data input was a DEM derived from digital photogrammetric means, in a raster data structure, which could be updated as new aerial photography becomes available.

The datasets of land cover visibility were compiled into a single dataset (a 'stack') in ESRI GRID format (ESRI, 1998) for interpretation within that software package and ERDAS IMAGINE (ERDAS, 1999).

4 Results

4.1 Assessment of change in land cover visibility

The changes in land cover contained in Table 1 show an increase in woodland from 130.5 ha to 518.1 ha (297%) and an increase from 75.2 ha to 156.0 ha (107.4%) in land interpreted as having woodland as a dominant component of a mosaic of different land cover types. The increase in the visibility of woodland over the same period has been from 2 639 ha to 3 329 ha, an increase in areal extent of 26%.

To assess the effect of the extent of the changes in plantation woodland, both the 'Woodland' class and the 'Woodland Mosaic (woodland dominant)' class were considered together. They show a combined increase of 468.4 ha in area of land cover, with a corresponding increase of 1 219.2 ha in the area from which such land cover would be visible. The extent of the change in area from which plantation woodland would be visible can be interpreted from Plate 3 (a) which shows the extent in 1946 and Plate 3 (b), which shows the extent in 1988, each with the outlines of the land cover interpretation of plantation woodland overlaid.

The geographic distribution of the visibility of land upon which plantation woodland was present in 1946 was greatest in the centre of the valley, whereas by 1988, this area had extended to cover almost all of the valley. A comparison of the two images suggests that a significant effect of the increase in the area of plantation woodland was in the amount that is visible from across the whole area of the valley, particularly on the hillslopes to the east. Figure 2 shows the difference in the extent of the visibility of plantation woodland, overlaid on the orthophotograph of the area for 1988 for reference.

The high plateau in the south of the area shows no change in visibility of plantation woodland, with an area containing no views of plantation woodland at either date. The land from which views of plantation woodland have decreased is concentrated in two areas, the valley bottom in the north of the area, and the south east corner where there was an interpretation of a block of broadleaf woodland from the 1946 aerial photography, but this area been interpreted as being open, scattered, woodland in 1988. This change in interpreted land cover has led to a significant change in the derived visibility of plantation woodland. Such a change may be a reflection of grazing pressures, causing a reduction in the density of the woodland cover (Hester *et al.*, 1996); thus the derived visibility of open woodland would show a commensurate increase. However, it should be recognized that there is also the potential for error in the aerial photographic interpretation, which could lead to the same outcome, but the consequences for the subsequent analysis and interpretation can be highly misleading.

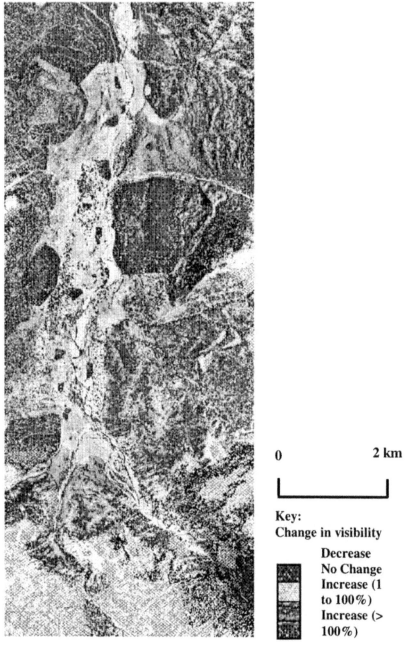

0   2 km

Key:
Change in visibility

Decrease
No Change
Increase (1
to 100%)
Increase (>
100%)

Figure 2. *Change in visibility of plantation woodland between 1946 and 1988, superimposed upon orthophotograph for 1988*

4.2 Changes in combinations of land cover in a view

The inventories of land cover visibility for each year were combined to produce an output that represents the number of land cover types visible from each cell, and the proportions of the total view from each cell occupied by each land cover type. This provided a basis

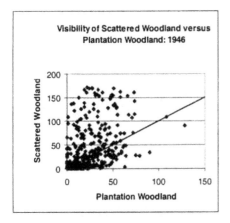
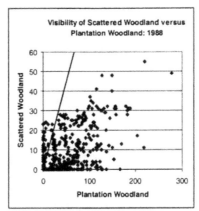

Figure 3 (a). *Distribution of visibility of land from which both scattered woodland and plantation woodland are visible, for 1946. (1:1 line shown for information).*

Figure 3 (b). *Distribution of visibility of land from which scattered woodland and plantation woodland are visible, for 1988. (1:1 line shown for information).*

for assessing the combinations of land cover types visible to an observer, which in future could be used in modelling landscape preference (Shafer and Bush, 1977; Wherrett, 1998).

Figures 3 (a) and 3 (b) show plots of the visibility of woodland that is scattered or mosaiced with non-woodland understorey vegetation, compared to that of plantation woodland. The graphs show an increase in the area from which the plantation woodland was visible and a drop in the area from which the scattered woodland was visible. This is as a direct consequence of the felling of some of the low density woodland and replanting with commercial, coniferous, forestry. The impact upon the content of landscape views is that 54.7% of land with both cover types present had a greater proportion of scattered woodland than plantation woodland in 1946, compared to only 6.1% in 1988.

4.3 Diversity in visible land use

Figure 4 shows the spatial distribution of diversity in land use visibility for each cell, coded in terms of the number of land use classes that were visible for each year. The Figure shows, that fewer land use classes are visible on the higher land around the valley, compared to the valley sides, from which up to twelve classes are visible in 1946 (Figure 4(a)), and in 1988 (Figure 4(b)). In general, increases in diversity of visible land uses between 1946 and 1988 are predominantly in the southern part of the area, whereas decreases have occurred in the northern part of the area. The decrease can be explained, in part, by the increase in plantation woodland in the north west and central parts of the study area.

To illustrate how the area from which different land use types may be visible can be used in assessing the potential impact upon visitor experience of the landscape, an example of the measurement with respect to footpaths is shown in Figure 5.

A route, going approximately from west to east across the glen, has been compiled from tracks that were interpreted from the orthophotographs for 1946 and 1988, and this route has been used to derive a profile across the combined surfaces of visibility for each

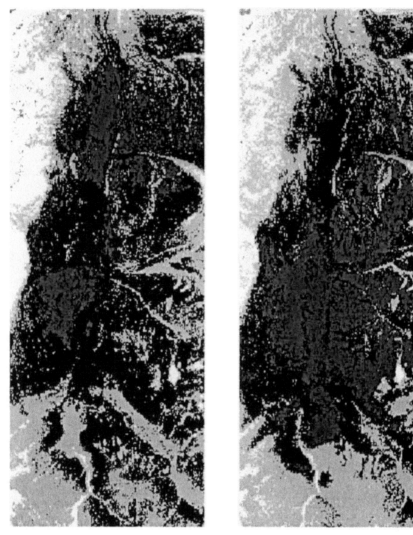

Number of Land Use Classes	Class Colour
1 - 3	
4 - 6	
7 - 9	
10 - 12	

Figure 4. *Diversity in land uses visible from each cell in (a) 1946; (b) 1988.*

year. The route is shown in Figure 5 (a), superimposed upon the orthophotography for 1988. Figures 5 (b) and (c) show the extent of the visibility of plantation woodland along the route of the pathway. The woodland in 1988 is visible from both a greater length of the route, and to a greater extent, than in 1946. The point from which most woodland was visible in 1988 is shown at location A on Figure 5 (a), which is a point on the hillside

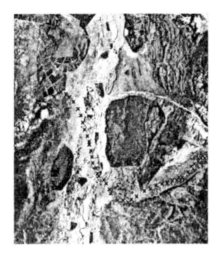

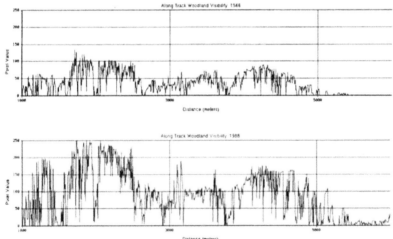

Figure 5. *Levels of visibility along a track, running approximately east to west across the valley. (a) Route of path, superimposed on orthophotograph for 1988; (b) Along track visibility of woodland; (c) Along track visibility of woodland.*

opposite which there has been extensive replanting over the period between the dates of the photographs.

The increase in plantation woodland has had a quantifiable impact on the views from the path. However, the significance of the change on users of this route cannot be directly inferred from the magnitude or geographical distribution of these changes for two reasons. Firstly, there are only two dates for which land use change has been interpreted and there is no information on the rate of change of land use, woodland growth, or woodland visibility over the 42 year period. Therefore, a recognition, or understanding, by observers that there is any visible change in the land use cannot be assessed. Secondly, the perception of the users of the track, of the content of the view will vary with respect to a number of factors, including atmospheric conditions, their attitude towards the function of the landscape (*i.e.* the relative benefits of production, diversity and aesthetics Daniel, 1999), and individual preference (Wherrett, 1998).

5   Discussion

The method described in this paper has provided one means of describing the visual resources of the study area, and the consequences of changes in land cover on those resources. The distinctive changes in the extent of both actual woodland cover and the visibility of woodland cover suggest that the content of the visual landscape has changed considerably over the period between 1946 and 1988. The changes in the composition of views, for example, those containing plantation and scattered woodland, would support the needs of the National Park objective of managing land to be 'sensitive to the design of man-made features', which includes plantation forestry. However, the method described does not provide an evaluation of the significance of these changes to the observer. Therefore, the changes in the spatial distribution of the land cover have not been assessed in terms of the potential impact on the elements that may contribute to peoples' preference for a particular landscape view.

The underlying spatial data used in any study such as that presented in this paper is subject to inaccuracy in its collection and representation. As mentioned previously, the methods described rely to a considerable extent upon the quality of the DEM and land cover data. Errors in the terrain models will impact on the intervisibility of locations, resulting in some estimates of landscape visibility being higher, or lower, than those of the initial calculation. Further methodological development is required to assess the sensitivity of the inventories of visibility to such errors, and the selection of radius of view, resolution of the input data and sensitivity to the heights of land cover features (such as forest canopies). An understanding of the inaccuracies in the underlying land cover data (MLURI, 1993; Aspinall and Pearson, 1995) is also now required to interpret the potential significance of this source of land cover error on the assessments of the visual resources of the landscape.

Three further limitations of the land cover data used as an input to this study are:

1.   The structure, composition and degree of homogeneity of the land cover classes that were used will be different for each geographical unit. For example, the actual density of trees within two polygons that were interpreted and coded as being the same class, will vary.
2.   Mosaics of two land cover types will increase the degree of difference between apparently similar polygons.
3.   The taxonomy of the land cover classification will be significant in the interpretation, and value, of the results.

Although the use of the interpreted land cover data has a number of limitations, it does provide a basis for planning at a strategic level, across a wide area. Alternative approaches to refining information on land cover to be considered in future include the identification, or recognition, of features from simulations of the landscape, either in the form of automated classifications of aerial photography or high-resolution satellite imagery, and the direct analyses of three-dimensional models, in which features are explicitly represented by models (Bishop and Gimblett, in press).

The efficiency of the computational aspects of the approach taken may be improved upon, but the balance between data resolution, CPU and lapsed computing time all require to be assessed with respect to the value and use of the output. The use of a raster data structure has an advantage of use within a wide range of commercially available software packages, with options available to implement the routines used directly within such packages. The length of lapsed time may be a limitation in the frequent use of the approach to deriving surfaces of terrain visibility, but such analyses are not required on a real-time basis, and form part of the data on the natural resource audits of an area, and can be treated in a similar manner to the collection of high resolution terrain or land cover data.

6   Conclusions

The methods and analyses presented in this paper provide one basis for assessing the spatial significance of changes in land cover in terms of the visual resources of the landscape. The application of these methods to the planning and management of the visual resources of the landscape previously been predominantly in the assessment of the potential visual impact of man-made structures, such as wind turbines. However, when applied to natural or semi-natural vegetation, they also have a potential role in the development of change in visual landscape resources.

The derived datasets also provide a basis for assessing the diversity of view content within an area, and current work is being done to assess the spatial relationships between the visibility of different land cover types, with a view to testing an indicator of the impact of changes in land cover or use on the visual landscape, that is sensitive to the shape, topographic context and nature of change.

Acknowledgements

The author would like to acknowledge the Scottish Executive Rural Affairs Department for their support of the research, and Scottish Natural Heritage and the Cairngorms Partnership for access to the aerial photography. Thanks are also due to an anonymous referee for their comments.

References

Aspinall, R.J. and Pearson, D.M., 1995. Describing and managing the uncertainty of categorical maps in GIS. In: P. Fisher (Editor) Innovations in GIS 2, pp. 71 – 83.

Bell, S., 1999. Landscape: Patterns, Perception and Process. E&FN Spon, London, pp. 288.

Benedikt, M.L., 1979. To take hold of space: isovists and isovist fields. Environment and Planning B, 6: 47 - 65.

Bishop, I. D., and Gimblett, H.R., Management of recreational areas: geographic information systems, autonomous agents and virtual reality. Environment and Planning B: Planning and Design, in press.

Brabyn, L., 1996. Landscape classification using GIS and national digital databases, Landscape Research, 21(3): 277 - 300.

Brown, T. C. and Daniel, T. C., 1987. Context effects in perceived environmental quality assessment: scene selection and landscape quality ratings. Journal of Environmental Psychology, 13(4): 283 - 291.

Coeterier, J. F., 1994. Cues for the perception of the size of space in landscapes. Journal of Environmental Management, 42(4): 333 - 348.

Cairngorms Partnership, 1999. The Vision for the Future. Cairngorms Partnership. pp 61.

Daniel, T.C., 1999. Whither scenic beauty? Visual landscape quality assessment in the 21st Century. In: E. Lange and I. Bishop (Editors). Proceedings of Our Visual Landscape, Ascona, August 1999.

DeFloriani, L. and Magillo, P., 1994. Visibility algorithms on triangulated digital terrain models. International Journal of Geographic Information Systems, 8(1): 13 - 41.

DeFloriani, L. and Magillo, P., 1999. Intervisibility on terrains. In: Geographical Information Systems, Second Edition (Eds. P.A.Longley, M.F. Goodchild, D.J.Maguire and D.W.Rhind), John Wiley and Sons, New York, Vol. 1, Chapter 38: 543-556.

ERDAS IMAGINE, 1999. IMAGINE 8.4 Users Guide. ERDAS , Atlanta, USA.

ESRI 1998. GRID Users Guide, ESRI, Redlands, Calif.

ESRI 2000. Using ArcToolbox: ArcInfo 8, ESRI, Redlands, Calif. pp 99.

Fisher, P.F., 1994. Probable and fuzzy models of the viewshed operation. In: M. Warboys (Editor). Innovations in GIS 1 Taylor and Francis, London, pp. 161-175.

Fisher, P.F., 1996. Extending the applicability of viewsheds in landscape planning. Photogrammetric Engineering and Remote Sensing, 62(11): 1297-1302.

Franklin, W. R. and Ray, C. K., 1994, Higher isn't necessarily better: visibility algorithms and experiments, In: Proceedings of the 6th International Symposium on Spatial Data Handling, Edinburgh, UK:751 - 770.

Gourlay, D. and Slee, W., 1998. Public preference for landscape features: a case study of two Scottish Environmentally Sensitive Areas. Journal of Rural Studies, 14: 248 - 263.

Hester, A. J., Miller, D. R. and Towers, W. 1996. Landscape-scale vegetation change in the Cairngorms, Scotland, 1946-1988: implications for land management. Biological Conservation, 77: 41-51.

Kidner, D.B., Rallings, P.J. and Ware, J.A. 1997, Parallel processing for terrain analysis in GIS: visibility a case study, Geoinformatics1(2):183 – 207.

Kliskey, A. D., Hoogsteden, C. C. and Morgan, R. K., 1994. The application of spatial-perceptual wilderness mapping to protected areas Management in New Zealand. Journal of Environmental Planning and Management, 37(4): 431 - 445.

Miller, D.R., 1999. Using aerial photography in static and dynamic landscape visualization, In: M. Usher (Editor). Landscape Character: Perspectives on Management and Change. The Stationary Office, Edinburgh. pp. 101 – 111.

Miller, D.R. and Law, A.N.R, 1997. Mapping terrain visibility. The Cartographic Journal, 34(2): 87-91.

Mladenoff, D.J., and Baker, W.L., 1999. Development of forest and landscape modeling approaches, In: D.J. Mladenoff and W.L. Baker (Editors). Spatial Modeling of Forest Landscape Change, Cambridge University Press, Cambridge. pp. 1-13.

MLURI, 1991. The Measurement and Analysis of Land Cover Changes in the Cairngorms, A report to the Scottish Office Environment Department. pp. 83.

MLURI, 1993. The Land Cover of Scotland: Final Report, A report for the Scottish Office Environment Department, Macaulay Land Use Research Institute, Craigiebuckler, Aberdeen.

Ordnance Survey, 1997. Profile User's Guide, Ordnance Survey, Southampton.

Peterken, G.F., Baldcock, D. and Hampson, A., 1995. A Forest Habitat Network for Scotland. Scottish Natural Heritage Research, Survey and Monitoring Report No. 44. Scottish Natural Heritage, Perth. pp. 119.

Shafer, E.L., and Bush, R.O., 1977. How to measure preferences for photographs of natural landscapes. Landscape Planning, 4, 237-256.

Steinitz, C., 1990. Toward a sustainable landscape with high visual preference and high ecological integrity: the loop road in Acadia National Park, USA. Landscape and Urban Planning, 19: 213 - 250.

Tandy, C., 1971. The 'isovist' method of visual analysis of landscape, Land Use Consultants, Croydon.

Teng, Y.A., Menthon, D. de and Davis, L.S., 1997. Parallelizing an algorithm for visibility on polyhedral terrain. International Journal of Computational Geometry and Applications, 7:75 – 78.

Wascher, D.M, Mugica, M and Gulink, H., 1999. Establishing targets to assess agricultural impacts on European landscapes, In: Environmental Indicators and Agricultural Policy (Eds. F. Brouwer and B. Crabtree), CABI Publishing, Wallingford, pp. 73 – 87.

Wherrett, J.R., 1998. Natural Landscape Scenic Preference: Techniques for Evaluation and Simulation. Ph.D. Thesis, Robert Gordon University, Aberdeen, Scotland.

Section 4

Applying the information: Decision Support.

14

User Interface Considerations To Facilitate Negotiations In Collaborative Spatial Decision Support Systems

Jaishree Beedasy, Rameshsharma Ramloll

In this paper, we describe the founding phases in the design of a collaborative spatial decision support system, Collaborative Spatial Multicriteria Evaluator (CoSpaME). Our approach is influenced to a large extent by the existing body of research in Computer Supported Cooperative Work (CSCW). CSCW researchers study collaborative activities in the real world and design collaboration enabling technologies. There is a broad consensus in this community about the essential role that negotiations play in the coordination of work activities in collaborative settings. We argue that when designing collaborative SDSSs, significant design effort has to be focussed on supporting negotiations. This is to be done in addition to interactive data visualisations which have naturally been of primary interest to GIS researchers. We illustrate our strategy with reference to a prototypical CoSpaME for illustrating the potential impact that simple interface considerations have on facilitating negotiations among peer decision makers.

1. Introduction

An increasing number of GIS researchers have shown interest in developing systems which facilitate solving semi-structured and unstructured problems defined by multiple and conflicting objectives and criteria (Jankowski 1997). Such systems typically allow the representation of complex spatial structures and include tools for spatial, geographical, and statistical analyses (Cooke 1992; Densham 1991; Crossland et al. 1995). We have developed a prototypical Spatial Decision Support System (SDSS), Spatial Multicriteria Evaluator (SpaME), which uses MCE techniques to help the siting of tourism resorts in Mauritius (Beedasy et al 2000). These are namely, Boolean combination, compensatory weighted combination [Saaty's AHP technique (1980) has been used for the development of criteria weights] and fuzzy combination using OWA operators (Yager 1987, O'Hagan 1988). A windfall benefit of the SpaME environment is that it allows the individual user to refine her understanding of the decision problem in terms of the relevant criteria and the interactions between them.

Since decision making problems, such as the siting problem mentioned earlier, frequently involve multiple stakeholders with potentially divergent perspectives, methods that contribute toward consensus building are required (Janssen R et al 1990). Various approaches have already been proposed. Those which emphasise the formal mathematical modelling of group decision processes are well known (Kersten 1999). In our opinion, such approaches while being undeniably useful embody the implicit assumption that problem solving or decision-making problems can be deconstructed into clearly defined parts. Each of these parts can then be solved concurrently or sequentially in isolation before being synthesized into a solution. We argue that many if not most real world spatial decision problems are too complex to be simply divided and conquered due to continuous information changes, high degrees of uncertainty, intricate relations and hard to quantify individual stakeholder expertise, reliability or trust. Attempts to capture faithfully and completely all the defining aspects of collaborative decision making under a unified mathematical model are likely to produce complex and difficult to use collaborative SDSSs.

2. A CSCW perspective on facilitating negotiation in group decision making

In this paper, we will present our design strategies which are based on lessons learnt in Computer Supported Collaborative Work (CSCW) research to enhance SpaME so that it facilitates negotiation between stakeholders. Our approach is informed by meta-analyses of the current corpus of workplace studies derived from observations (Grinter 1997,1998) of how individuals collaborate and the role of technology in that collaboration. The main observations are: (1) individuals make some of their work visible to others, and monitor each other, (2) researchers have shown that collaborative work is dynamic and involves many channels of communication, (3) people construct and share the interpretations of the work in progress, (4) maintaining context supports long term collaboration or *encourages sustained collaborative engagement between stakeholders* and, most importantly, (5) work often deviates from the planned activity in order to accommodate situated action (Suchman 1987). The first four observations confirm that the discourse of negotiations which includes the passing, accepting and understanding of messages (Firth 1995) play a significant role in collaborative activities and the last observation reiterates our opinion that group decision processes are difficult to formalise. This is also supported by the view that discussions, negotiations and bargaining are *unavoidable* dimensions of decision-making (Carver 1991).

Our main concern is to investigate human computer interface strategies to facilitate negotiation in collaborative decision-making. In other words, we are interested in what, how and when information can be presented to decision makers at the user interface so that they are provided with opportunities to coordinate more effectively their interactions in order to enable a more fluid, efficient and fruitful group decision making process irrespective of the spatial and temporal attributes of the relevant parties.

Our main contributions include accounts of (1) the design and implementation of awareness widgets (Gutwin et al. 1996) that will provide opportunities for every stakeholder to inspect and visualise otherwise hidden group decision making processes, (2) the linkage of relaxed WYSIWIS (What You See Is What I See) stakeholder specific views/interfaces (Stefik et al. 1987) to enable controlled access to awareness information describing peer activities and, (3) a preliminary evaluation of our collaborative SDSS interface which will reveal whether our focus on supporting the discourse of negotiation processes has any significant effect on the group decision activity as a whole in various space time configurations (Ellis 1991).

3. Our Approach

CSCW research has reported a number of efforts turning essentially single-user software applications into collaborative ones. Such resulting software is known as collaborative transparent applications (Lauwers et al. 1990). This bottom-up approach to groupware design where bottom is an existing infrastructure designed primarily for single user interaction and up refers to the resulting collaborative software has been popular. This is most probably because of the understandable attempt to minimise design and implementation costs by using existing infrastructure. Evaluations of software resulting from this strategy have revealed that groupware requirements typically are such that it might be preferable to start designing groupware from scratch rather than spend a lot of energy and money to patch up an existing single application in order to make it collaborative. This movement gave rise to a new design strategy which could be referred to as a top-down strategy. Here top refers to the functionality at the interface and down to the underlying infrastructure.

3.1 Eliciting Requirements for a Collaborative Version of SpaME

We adopt a top-down design strategy to design CoSpaME. To begin, we create a minimal CSDM system using SpaME and Microsoft NetMeeting™. This use of SpaME in conjunction with NetMeeting™ has facilitated our task of explaining to a group of potential collaborative SDSS users who had no experience in dealing with collaborative software of the possible functionalities of such an application. Four participants each with a different set of expertise and interests were asked to use our minimal NetMeeting-SpaME environment (Figure 1). They were given a siting problem to collaborate on while being dispersed over a LAN. They were then interviewed to gain their views on (1) the aspects of the collaborative environment that they thought facilitate collaboration, (2) those that actually hinder such a group activity, and lastly, (3) any suggestions they have to improve the minimal SpaME-NetMeeting™ collaborative environment. We then attempt to use whatever information gathered from this exercise to build a prototypical interface for CoSpaME in order to address their concerns as far as possible. This approach allows us to produce a framework that will drive the top-down approach for the development of CoSpaME.

We present here a summary of the output of this first requirement elicitation phase in the design of CoSpaME.

3.2 The main limitations of the SpaME-NetMeeting™ environment identified by the stakeholders are:

1. The SpaME-NetMeeting environment does not allow participants to work synchronously. Only one participant can actively use SpaME at one time while her peers are mere observers. This limitation imposes an artificial social protocol that drives the order in which participants can use SpaME. This often leads to situations

where participants are fighting for floor control or participants being perceived as being privileged users of the collaborative environment.

2. Even when the stakeholders are co-located, that is in addition to situations where they were spatially distributed, they express difficulty in being continuously aware of the dynamic negotiation status of their peers so that the planning and coordination of their negotiation moves is not facilitated.

3. As in most real world scenarios, stakeholders find it difficult to trust some individual perspectives. This is especially true if the formal group decision model which is supposed to have synthesised a best solution from internal workings of the model are hidden and difficult to understand by non-specialists.

4. Most participants express concern that all their interactions with SpaME are public and that they constantly have to face evaluation apprehension stress. The fact that all their activities at the interface are public also discourages them to perform and explore bold what-if scenarios.

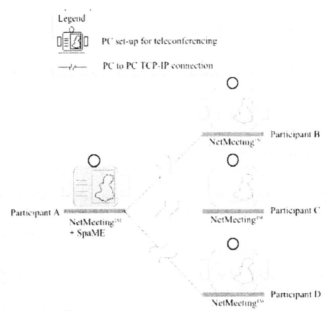

Figure 1 *The NetMeeting^{TM}-SpaME collaborative set-up*

3.3 The aspects of the SpaME-NetMeeting^{TM} environment identified as enablers of collaborative interactions are:

1. All participants praise the environment for providing them with the opportunity for full duplex audio and video conferencing. Ironically, much of this medium of communication has been used to resolve conflicts arising in the breakdown of the social protocol to have access to SpaME.

2. Participants agree that the MCE techniques provided by SpaME are useful in helping them to tackle their spatial decision making more rapidly taking into consideration many more criteria than would otherwise have been possible. This issue has been discussed in detail elsewhere (Beedasy et al 2000).

3. The users who are impacted by such spatial decisions feel that they can contribute to the decision making process in such a collaborative environment.

4. The participants concede that such settings can bring better consensus in decision-making and lead to more harmonious 'outcomes' especially as siting problems involve conflicting objectives.

3.4 The facilities that participants propose to improve the collaborative set-up are:

1. Participants believe that the time taken to complete a group decision process in the current SDSS, which enforces a strict turn-taking strategy for interaction, is much higher than if the stakeholders are allowed to interact with SpaME synchronously.

2. Stakeholders express the need to have full control on the amount of local information to be provided to peers. This privacy control issue has emerged as a central factor that will determine whether stakeholders are really going to use the collaborative decision making software in a real-world setting.

3. Participants feel a lack of appropriate tools that will allow them to annotate the spatial displays both textually and graphically and to share these annotations. Such facilities are essential both as an external memory for the collaborative decision making process and as a medium for constructing complex arguments.

4. Stakeholders mention that they need better facilities to analyse the output of their decision-making processes rapidly. This is partly because they are eager to determine the extent to which the collaborative solution deviates from their own throughout the collaborative decision-making process.

5. Participants also request for opportunities to have access to information about peer activities that can potentially prove useful in enabling fruitful interventions or interactions in a collaborative setting.

6. Two users also express their preference to have access to multimedia information - real photos or videos of the actual sites that demand negotiations.

3.5 A user view of prototypical CoSpaME interface

We have designed a series of prototypical CoSpaME interfaces in MacromediaFlash™ to address as much as possible the concerns of the stakeholders mentioned earlier. Half a dozen of design iterative steps have produced the prototypical CoSpaME interface which is satisfactory to the four different stakeholders. The spatial reporting capability and visualisation representation is still being investigated. In this prototypical version, chorognostic displays (Armstrong et al 1992), which provide users with geophysical information about the study area, are provided. Multiple maps have been considered for viewing results of different participants simultaneously.

The results analyser allows participants to use direct manipulation to construct Boolean queries on peer participant results. These can be viewed in the Spatial Views window (Figure 2). Facilities for reporting non-spatial information using tables and graphs are also being planned. The current features of the interface shown in figure 2 are described in Table1. A demo of CoSpaME is available for download at www.dcs.gla.ac.uk/~ramesh .

Feature number	Description
1	Annotation tools which can be used by the participants on the Spatial Views window when maps are displayed.
2	Participant targeted by reveal-spotlight
3	Spatial Views window where the maps and results are displayed
4	Spotlight which reveals to Participant A the features which participant B (target) is pointing at on another map at any time.
5	Display of selected chorognostic (Armstrong et al 1992) or results maps or query outcomes
6, 7	Privacy controls (varying from high to low)- Applying the simple reciprocal mutual awareness rule e.g. in high privacy mode participant's moves are not revealed to peers and peers' moves are not revealed to participant
8,9	Sample of chorognostic maps relevant to the session
10,11	Peercams showing videos of the users who are currently participating and are not in private mode.
12	Result analyser consists of a Venn diagram that allow rapid queries of the results using single clicks of the diagram. Query outcomes are displayed in the Spatial Views window(3) on the right hand side. When the name of a participant is clicked on the result analyser the results based only on the weights of that single user is displayed.
13	The values chosen by the participants (on the nine point scale) when a pair of criteria is compared shown in the respective colour codes of each peer.
14, 15	Specific criterion – (criteria represented as a matrix for easy comparison)
16	The nine-point scale used to compare each pair of criteria by clicking on the appropriate point on the scale
17	Pairwise comparison of the criteria for calculation of weights (AHP technique)
18	Radar Views showing potential regions of peer interest based on their cursor position
19	Region around cursor position for specific user with top left inlay showing a video of that participant
20	Current cursor position of each participant (in their respective colour codes) on a map

Table 1 - *Collaborative GIS interface widgets*

4. Discussions And Future Work

This short paper has focused on the initial requirements capture phase for the design of CoSpaME and on a description of a prototypical CoSpaME interface that they directly inform. The direct involvement of prospective users of our collaborative software has allowed us to produce an interface that is likely to match their needs. Many of the innovative widgets that have been described are the results of interesting debates that the SpaME-NetMeeting[TM] arrangement gave rise to. We therefore thank the participants for their contributions to the ideas presented in this paper. The prototypical interface has produced a framework that will implicitly inform (1) the choice of existing collaborative software development toolkits needed for the implementation of the full blown system (2) and the underlying software architecture of the application. Currently, we are considering the Java Shared Data API , Habaneros and Java Collaborator Environment as prospective candidates for suitable collaborative software development environments.

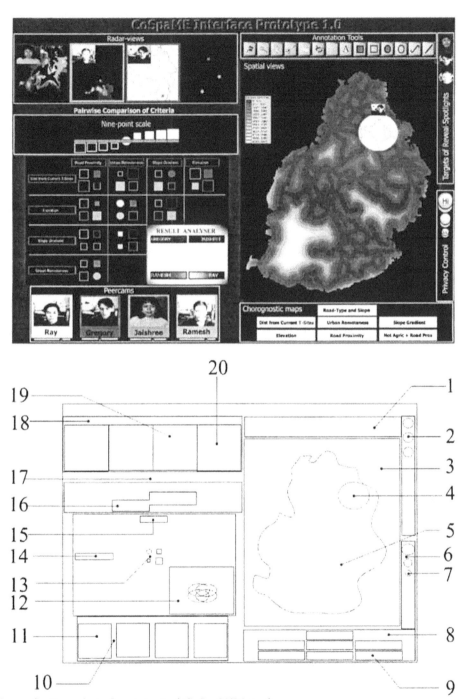

Figure 2 - *Snapshot of prototypical CoSpaME Interface*

On a final note, the current version of the prototype has been designed with only 4 participants in mind. There are obvious issues that need to be considered when the number of participants scale up. However, whether synchronous collaborative interactions of the degree of complexity described in the paper occur for larger groups of stakeholders is an issue to be investigated. We believe that for larger collaboration scenarios such as that which will involve a public GIS will call for another requirements capturing phase on its own.

References

Armstrong, M.P., Densham, P.J., Lolonis, P., Rushton(1992). Cartographic displays to support locational Decision Making. *Cartography and Geographic Information Systems*, Vol. **19**, No.3,:154-164.

Beedasy, J. and Whyatt, D. (2000) Spatial decision support system for tourism planning in a developing island' to appear in *International Journal of Applied Earth Observation and Geoinformation*, ITC, Netherlands.

Carver, S. (1991) Integrating multicriteria evaluation with GIS, *International Journal of Geographical Information Systems* **5**(3), 321-339

Cooke, D.F. (1992) Spatial Decision Support System: Not Just Another GIS. *GeoInfo Systems*, **2:5**, 46-49.

Crossland, M.D., Perkins, W.C., and Wynne, B.E. (1995) Spatial Decision Support Systems: An Overview of Technology and a Test of Efficacy. *Decision Support Systems*, **14**, 219-235.

Densham, P.J. (1991). Spatial Decision Support Systems. In: *Geographic Information Systems: Principles and Applications*, Vol. 1, edited by D.J. Maguire, M.F. Goodchild, and D.W. Rhind, pp. 403-412. London: Longman Scientific and Technical.

Ellis C.A., Gibbs S.J. and Rein, G.L. (1991): ``Groupware : Some issues and experiences". *Communications of the ACM*, vol.**34**, no.1, 1991, pp. 38-58.

Firth, A. (ed.), (1995), The Discourse of Negotiation. *Studies of Language in the Workplace*, New York: Elsevier.

Grinter, R. E. (1997): From Workplace To Development: What Have We Learned So Far and Where Do We Go? , *Proceedings of the International ACM SIGGROUP Conference on Supporting Group Work (GROUP'97)*, Phoenix, Arizona, November 16-19, pages 231-238.

Grinter, R. E. (1998): Recomposition: Putting it All Back Together Again, *Proceedings of the Computer Supported Co-operative Work (CSCW'98)*, Seattle, Washington, November 14-18, pages 393-402.

Gutwin C., S. Greenberg, and M. Roseman (1996). A Usability Study of Awareness Widgets in a Shared Workspace Groupware System. *In Computer-Supported Cooperative Work*, pages 258-67. ACM Press, 1996.

Jankowski, P., Nyerges, T.L., Smith, A., Moore, T.J., Horvath, E. (1997) Spatial Group Choice: A SDSS tool for collaborative spatial decision making, *International Journal of Geographical Information Systems* **11**(6) Pp566-602.

Janssen R , Rietveld P (1990) Multicriteria Analysis and GIS : An application to agricultural land use in the Netherlands in *Geographical information systems for urban and regional planning*. Eds Scholten H J and Stillwell J.C.H P129-139.

Kersten, G. (1999) *Support for Group Decisions and Negotiations* available on-line at http://www.iiasa.ac.at/Research/DAS/interneg/research/misc/intro_gdn.html

Lauwers, J. C. and K. A. Lantz (1990). Collaboration Awareness in Support of Collaboration Transparency: Requirements for the Next Generation of Shared Window Systems. *Proceedings of ACM CHI'90 Conference on Human Factors in Computing Systems:* 303-311.

O'Hagan M, (1988) Fuzzy Decision Aids. *Proceedings of the 21st Asilomar Conference on Signals, Systems and Computers*

Saaty T.L (1980) *The Analytic Hierarchy Process* RWS Publications

Stefik M., D. G. Bobrow, G. Foster, S. Lanning, and D. Tatar (1987). WYSIWIS Revised: Early Experiences with Multiuser Interfaces. *In ACM Transactions on Office Information Systems*, pages 147-167. ACM Press, April 1987.

Suchman, L. A (1987*): Plans and Situated Actions: The problem of human-machine communication.* Cambridge University Press, 1987.

Yager, R.R (1988) On ordered weighted averaging aggregation operators in multicriteria decision making . *IEEE Transactions on Systems, Man and Cybernetics* Vol **18** (1), 183-190

15

The Use Of GIS In Brownfield Redevelopment

Rebekah Boott, Mordechai Haklay, Kate Heppell, Jeremy Morley

In recent years, the issue of Brownfield site development - the re-use of previously used urban land - has gained a significant place in the planning agenda. However, not all Brownfield sites are derelict or contaminated land, some are significant as environmental amenities - be it part of wider ecosystem or a green area for the local population. The growing concern to include environmental aspects into the public debate have lead the Environment Agency, the Jackson Environment Institute and the Centre for Advanced Spatial Analysis to commission a short term pilot study to evaluate the contribution of a GIS for decision support and for "discussion support".

In this paper, we describe how the state-of-the-art in geographic information (GI) and GI Science (GISc) can be used in a short term and limited project to achieve a practical and usable system. We are drawing on developments in information availability, as made accessible through the World Wide Web and research themes in GISc ranging from Multimedia GIS to Public Participation GIS.

1. Introduction

To meet the growing demand for housing it is forecast that an increase in 4.4 million homes will be required in the UK by 2016 (Urban Task Force, 1999). It has been proposed that brownfield[1] redevelopment would provide the space needed for the extra housing.

Brownfield sites are an important component of the government's strategy to develop sustainable urban communities. Sustainable development incorporates economic, social and environmental needs. It is concerned with reconciling economic demands and social needs with the capacity of the environment to cope with pollution and to support human and other life (Environment Agency, 1998). However, there is a lack of integrated and comprehensive knowledge on the condition, location and management of brownfield sites throughout the UK and little attention has been given to the contribution that brownfield

---

[1] For the purpose of this research the term 'brownfield site' was defined as an area which had previously been developed, but had since fallen derelict and may, or may not have existing buildings on it. No assumptions were made regarding contamination or pollution of each site.

sites make to biodiversity, nature conservation and amenity. The project described in this paper, funded by both the Environment Agency and University College London, aimed to evaluate the contribution of GIS to help remedy these issues.

It is in this context, that the National Land-Use Database (NLUD) should be mentioned. NLUD (as the acronym clearly states) is aimed to provide a detailed geographical record of land use in England and developed as a partnership between the Department of Environment, Transport and Regions, English Partnerships, Local Government Management Board and the Ordnance Survey (OS). Under current development and policy pressures, the partnership focuses on the collection and registration of brownfield related information (NLUD, 2000). However, although it is geographic, this system does not hold any information in a GIS format at the current stage (though plans for such data collection exist). Furthermore, it is very difficult to integrate the NLUD point data with geo-referenced data sets such as OS Landline data. Without ignoring the value and importance of NLUD, we decided to focus on GIS oriented data collection and assembly.

It was decided that we would focus on urban brownfield sites in the Wandle Valley, South London as these have the most redevelopment potential and the greatest economic, environmental and social benefits. It was envisaged that such a system would provide a tool that can be used by property developers, planning professionals, local authorities, environmental bodies and organisation and last but not least - the local population. Such a system can be used by the developer and local authorities to facilitate investment in urban areas by identifying sites with appropriate potential for re-development. Environmental professionals can use it to evaluate the risks from pollution of land, air and water in the context of such projects. Low income and minority populations' interests can be benefited from such a system by redirecting investment and revitalising their neighbourhoods. Finally, it can be used to encourage participation of various citizen groups in the planning process and balance the demands of development with the need to protect and enhance the environment.

Though we identified a wide range of stakeholders that are relevant to brownfield site redevelopment, we have not approached the system development with the idea that a single closed system could satisfy all. From the outset, it was clear that we would have to confront the opposing and what might seem as incommensurable views of different stakeholders. Nevertheless, it was felt that common information needs could be identified and accommodated. Furthermore, as the Environment Agency was a major stakeholder our main focus was on the environmental aspects of brownfield development.

It was decided that the system would be developed rapidly using PC based GIS mounted on a notebook computer. This was due to the strong presentation element in the project. As the project was developed under severe time and resource limitations, we decided to rely on existing digital data sets as much as possible and to combine recent lessons from multiple areas of Geographical Information Science (GISc) research. This paper follows the development of the Brownfield GIS, pointing to the areas that informed it and the way in which current development in information access and Geographical Information (GI) were used to develop the system.

The paper starts with the identification of user requirement. We have used multiple approaches to identify and accommodate the requirements and needs of multiple stakeholders. Based on those requirements, we have built the system - by that we mean mainly the data collection and database organisation. As will be explained latter, it was felt that the system should be left as "open-ended" as possible; therefore the interface of the hosting GIS software (ArcView) was left relatively unchanged. In the following section we discuss several analyses that have been carried out with the system and the few customisations to the systems' interface that were deemed necessary. Following this description, we describe the exposition of the system to the different stakeholders, starting with a local activists who participated in a workshop where they learned about GIS capabilities and uses and then continuing with the more institutional users, such as Environment Agency officials, local authorities GIS and planning officers and the Government Office for London. We then turn to discuss the main lesson learned from this project that are relevant to the broader GISc community. This includes the rapid development of GIS and its database, the contribution of GIS to the brownfield debate and the use of GIS as "discussion support" tool. We conclude with some recommendations for future research.

By its nature, the project is based on multiple research streams in GISc. We therefore felt that it is more appropriate to describe each of the area in the appropriate section of this paper and to combine them in the conclusion of this paper.

2. User Requirements

The integration of a user requirement study is now commonplace in general information systems design (Preece, 1995) and in GIS design (Reeve and Petch, 1999). They stem from studies in Usability Engineering and approaches like User Centred Design (Landauer, 1995) developed during the late 1980s. This aspect of GIS design received attention in the mid 1990s (Medyckyj-Scott and Hearnshaw, 1993; Nyerges *et al.*, 1995). However, in their more familiar form, user requirements are connected to "task analysis" - a process of identifying the tasks that the user performs with the system (see for example Rasmussen, 1995). This approach requires quite a high-level of knowledge about users' work practices, activities and the ways in which the proposed information system can support these processes.

In the context of the brownfield GIS and especially in light of the emphasis on environmental information, such clear tasks and processes are hard to define and any focus on specific process (such as site selection for property developers) will limit the usefulness of the system to other stakeholders. Therefore, we have selected a more integrative and deliberative approach, which combines methods like interviews with key stakeholders, a seminar series with academics and practitioners and the use of an inclusionary workshop. These methods stem from the latest ideas about the meaning of participatory planning (Healey, 1997; Healey, 1998). Furthermore, current functionality of desktop GIS contains the needed elements to accomplish a wide range of tasks. As such, they represent a "toolbox" that can be adapted to the specific task (Batty, 1993). Therefore, it was necessary to use the desktop GIS with its set of extensions as a test bed for ideas and to envisage the development of more closed and specific system latter on, when the main requirements and need will be clarified.

The user requirement study started with the identification of potential users, as was described earlier. Even though a wide range of stakeholders was identified, the core elements of the system were set and defined by the professional group. The rational behind this decision is that those are the most likely users of such system. Even this user group was heterogeneous. It includes national government officials from the Department of Environment, Transport and Regions (DETR) or Government Office for London (GOL), local governance officials from the up-coming Greater London Authority (GLA) or London Planning and Advisory Committee (LPAC) and London Boroughs. We have also considered Non-Governmental Organisations (NGO's), which are active in issues related to brownfield development.

Initially, the goals were defined through interviews with local and central government officials, representatives of the Environment Agency and others. These one-to-one interviews were augmented by views and ideas about brownfield development that came from a seminar series in UCL, held during 1999. These deliberative discussions focused around four themes (Bloomfield, forthcoming): First, the issues of *definition, criteria and survey* were raised. During this seminar, the question of "What is a brownfield site?" was examined. It became clear that though some general concepts exist, there is no exact definition that will suit all parties. The following seminar focused on d*ata, users and site assembly processes.* This seminar examined the information required by the various stakeholders. It clarified the difficulties that face developers, planners and others when they try to deal with brownfield sites. Such an operation includes the evaluation of the stock of sites or prioritises their development. The third seminar, titled *public involvement, skills and knowledge* considered the role of public participation and how should local and national interest groups should get involved in Brownfield site development. In the final seminar on *future government structure*, the focus turned to issues of governance and how bodies such as the GLA should act toward brownfield development These seminars participants came from a wide variety of backgrounds – including policy makers, officials, representative of the commercial sector and academics.

Though these seminars were not dedicated to the development of the brownfield GIS *per se*, we have used the opportunity to bring issues of information access and availability to the discussion table, as to improve out understanding of what was needed. It is important to note that during all seminars, participants raised the issue of information needs and GIS integration repeatedly and there was no need to "divert" the discussion to raise these issues. Many participants felt that the integration of GIS with public access medium (the Internet) would be very useful. Data availability was an important issue and it was suggested that improved availability/accessibility to information on the system might benefit the public. Participants also voiced concern about the creation of a system designed with the objective of only aiding developers in their search for sites to suit their own purpose. Many participants also wanted to see contextual as well as site-specific information. By including environmental or policy-based text on the system users would be able to view brownfield development as one component in the regeneration programme of a whole area. Regular updating of the system was also raised.

The seminar series and the interviews formed the first phase of the user requirement study. After this phase, the data sources for the system were gathered and the initial database created. These were followed by an evaluation of public requirement and need from such a system, as a workshop titled "Tools for planners, tools for the people?" revealed. This workshop is discussed later in the paper. First, to build a full picture of the database as it was used in the workshop, the process of data integration and analysis will be explained.

3. Data Collection

Traditionally, data collection and collation is considered as one of the complex and expensive task that relates to the creation of a GIS (Huxhold and Levinsohn, 1995). However, this situation is rapidly changing - at least for certain types of applications and activities[2]. The availability of digital data set and the ability of commercial-off-the-shelf (COTS) software to integrate them easily was predicted in the early 1990's by Batty (1993) and it can be argued that we have entered this era of easier and faster collation of GI data-sets.

However, some caveats do exist and will continue to do so. Though the problem of precision that stem from constraints in hardware or software - a problem that blighted earlier systems (Tomlinson, 1970) - has virtually disappeared, other issues like accuracy in GI (Burrough, 1986), conflation (Laurini and Thompson, 1992) or error propagation (Heuvelink, 1998) have not. While being aware of those issues and the possibility that specific questions cannot be answered without a complete, current and accurate brownfield database, for pragmatic reasons (such as the lack of information and the relative urgency in the requests from the user community), it was decided to trade-off

Name	Source	Description
Green/Natural environment Data		
Sites of Special Scientific Interest (SSSI's)	Environment Agency	Areas that have been designated by English Nature as being of outstanding value for their flora, fauna or geology under the Wildlife and Countryside Act 1981.
Metropolitan Open Land	Environment Agency	Areas within the built-up area that are a significant environmental resource to London.
Rivers	Environment Agency	This includes the Thames, Wandle and Beverley Brook rivers.
River Floodplains	Environment Agency	The limits of the floodplain are defined by the peak water level caused by rainfall of a 1 in 100 year return period so such a storm has a 1% chance of occurring in any particular year. The Environment Agency has a statutory responsibility for all flood defence matters concerning main rivers under the Water Resources Act (1991).
Sites of Metropolitan Importance	London Ecology Unit	These sites have the highest priority for protection and contain the best examples of London's habitats alongside rare species or assemblages of species or sites that have particular significance within large areas of heavily built-up London.
Sites of Local Importance	London Ecology Unit	These are sites of particular value to nearby residents or schools and are particularly important in areas otherwise deficient in nearby wildlife sites.
Sites of Borough Importance (I & II)	London Ecology Unit	These sites are important in a borough-wide view but have been split into two sub-categories on the basis of their quality. Damage to any of these sites would result in a significant loss to the borough.
River Thames	Ordnance survey	This file was produced from the Ordnance Survey Meridian data.
Archaeological Priority Zones	Environment Agency	These are areas known to be of archaeological importance because of past finds, excavations or historical evidence.

Table 1. *Data used in the Brownfield Project*

---

[2] And for certain organisational settings. It is important to remember that the brownfield GIS was developed inside a university and research group that already collated massive amount of GI, a fact that reduces the costs of GI quite dramatically.

Name	Source	Description
Infrastructure Data		
Meridian Data	Ordnance Survey	The Meridian dataset was used to show motorways, A roads, B roads and minor roads and mainline railway lines.
Landline Data	Ordnance Survey	This dataset was experimented with as a source of detailed local information.
Overland and Underground Stations	Own	Identifying a six-figure grid reference for each station from a 1:20,000 street atlas collected this data.
Socio/Economic Data		
Population per Enumeration District	Manchester Information & Associated Services (MIMAS)	Socio-economic data was provided through MIMAS
Main Shopping Areas	Unitary Development Plans (UDP's)	These are centres providing a range of facilities for the local population including shops, employment, social and community facilities, transport services, leisure and entertainment.
Brownfield Site Data		
Brownfield sites.	UDP's	These are areas designated by each of the four boroughs for redevelopment and range from vacant land to empty shop units.
Wandle Valley Regeneration Partnership (WVRP) Brownfield Sites.	WVRP	WVRP provided comprehensive information on nine key development sites in the Wandle Valley Strategic Employment Corridor. This data was in an analogue format and was digitised onto the system using ARC/INFO GIS. The sites are: • Cane Hill Park, Croydon • Former CMA site in Morden Road • Springfield Hospital, Tooting • Plough Lane Football Ground, Wimbledon • Beddington/Purley Way Cluster • Former Beddington Tip Site, Beddington Lane • Site North of Goat Road • Beddington Farmlands • Anchor Business Centre, Beddington Lane
National Land Use Database (NLUD) Brownfield Sites.	LB Sutton	This data was supplied so it could be compared to the Brownfield polygons digitised from the UDP maps.

Table 1 (continued). *Data used in the Brownfield Project*

high accuracy with information availability. Therefore, we have focused on the collation of data sets that can complement one another (especially in the environmental side). Another approach was to use known high-quality and up-to-date data sets, such as the Ordnance Survey (OS) data. Finally, we have tried to combine data sets that demonstrate the capabilities of GIS and spatial analysis in manipulation of socio-economic data (Martin, 1991) and environmental modelling (Goodchild *et al.*, 1993). These data sets came from many diverse sources in both digital and analogue formats. Table 1 summarises the data set that we used to compile the system database.

The datasets categorised as **green/natural environment** data in Table 1 were chosen to show how environmentally and culturally sensitive areas could influence the brownfield redevelopment process. The data acquired from the London Ecology Unit (LEU) provides important information for developers - as the development of some of the brownfield sites might be inhibited by environmental concerns. Noteworthy is that this information was digitised especially for this project and was not available in digital form previously. Flood plains information was based on the Environmental Agency modelling and was integrated in order to demonstrate how the output of GIS based environmental modelling can be integrated and used in the project context. By simple overlaying, it was possible to identify brownfield sites with a potential risk of flooding - an issue of importance for developer and planner alike.

The **infrastructure** datasets in Table 1 were chosen as they provided a good backdrop over which to view other datasets. OS Meridian data was chosen so that the road network could be used for service area analysis as we describe latter. Aerial photography data taken from Cities Revealed™ was integrated into the system to cover some of the main brownfield sites. This was used to provide a rich, contextual information for the brownfield site and its surroundings. The system was programmed in such way that once the theme is viewed at a greater scale than 1:10,000 the aerial photography data replaces the more skeletal road network while the other information relating to the brownfield sites remains visible. By doing so, it is possible to put the sites in better context and to help users to orientate themselves.

The population **socio-economic** data was included to show which areas could already have high-density population despite being brownfield sites. They provide an opportunity to juxtaposition relatively abstract information (population density per enumeration district) with physical or modelled information. However, we have limited our use of the census data set, as it was felt that it becoming dated and too inaccurate. The shopping area data was included so network analysis could be carried out to establish which brownfield sites were within the each shopping centre service area.

The **brownfield** data shown in Table 1 that was derived from Unitary Development Plan (UDP) maps was used because it was the most recently available. Originally we planned to use data collected for the National Land-use Database (NLUD) until its lack of availability and its difficulty for integration with GIS was discovered. Instead older data was digitised from the UDP maps and a sample of NLUD point data for Sutton was used in comparison. The Wandle Valley Regeneration Partnership (WVRP) brownfield site data was used as a direct result of the iterative development process. The nine sites were deemed of high importance to the partnership.

Together, the data sets represent a cross section of the issues that influence brownfield site development, with an emphasis on the integration of environmental data sets with socio-economic and infrastructure data sets.

4. Textual Data Collection And Multimedia

The integration of multimedia with GIS started during the mid 1990s (Craglia and Raper, 1995). Soon after, GIS vendors started to integrate multimedia capabilities into their products. However, such integration was somewhat awkward and the multimedia functionality usually limited. The introduction of the World Wide Web as a medium that is inherently multimedia based followed soon after and changed the way in which multimedia was distributed and organised. It is important to note that the environmental application of multimedia GIS have used the precursor of the Web environment - the acclaimed HyperCard (Fonseca *et al.*, 1995; Shiffer, 1995). However, the web browser, which is now part of any operating system makes the task of integrating multimedia into applications far easier. In some cases, the whole GIS application is integrated into the web browser, for example, Kingston *et al.* (2000) discussed such application that is more akin of traditional GIS, while others (Brown, 1999; Doyle *et al.*, 1998) discussed the potential of integrating other multimedia forms with Web-based GIS. In the case of the brownfield GIS, we used the capabilities of the web browser and the WWW to integrate multimedia and information held in remote servers to augment the capabilities of the basic software package, without "struggling" with the limited multimedia capabilities of the specific software.

We have created web pages to provide information on brownfield sites, ex-brownfield sites, rivers and areas of nature conservation. These pages include textual information and pictures taken with a digital camera for various sites. Links were established from many polygons in the GIS to the Web browser so that further information, which is available on the Internet, could be easily and quickly accessed. Such links were established to the Environment Agency pages, which deal with regulations and the agency responsibility areas. Another major source for external information was the Wandsworth council web site, which provides public access to its planning register and enables access the applications that relate to specific brownfield. Some brownfield sites also had web pages set up by local independent groups and we have linked to these sites. One such site is the Battersea power station site, in which links were establish to the various stakeholders - developers, local pressure group and the planning register.

5. GIS Analysis And Customisation

One of the most powerful features of any GIS is the capacity to carry out various spatial analyses quickly and easily. In the case of the brownfield GIS we chose to demonstrate these capabilities by implementing some overlay analysis - probably the most used analysis function of GIS since McHarg popularised it in the late 1960s (Mcharg, 1969). These analyses were integrated with outputs of network analysis and visualisation of service areas (Armstrong *et al.*, 1992). Noteworthy is the integration of the EA floodplain analysis as a given layer of information.

The green data sets were used (see Table 1) to show which brownfield sites were over 400 metres from parkland (using buffer analysis). It was thought these sites would be less desirable for housing because of the reduced easy access to green areas. A more sophisticated network analysis was carried out. For example, service areas of 1Km from mainline and underground stations and weighted service areas for shopping centres (according to the size of the centre) were identified and the results were used those brownfield sites that apply to these criteria. It was thought that this process could be used to identify which brownfield sites met a certain number of planning criteria such as less than 1Km from a shopping centre, but within 1Km of a tube station.

Even though the analysis was carried in a dense urban area, not all the service areas are circular in shaped - which hints to the inappropriateness of simple buffer analysis in this case. For ease of use and for visualisation purposes, various overlay analyses were carried out. For example, brownfield sites that fall within the boundaries of the calculated flood risk area were selected and stored as a separate group.

As was mentioned earlier, we tried to limit the customisation of the software. However, the few changes from the original, out-of-the-box functionality included the integration of a postcode based search tool. This was done to enable the use of an easy and familiar geographical reference, which is common and accessible to a wide range of users (Raper *et al.*, 1992). The second change was the connection to web based information browsing we have mentioned before. Finally, for the visualisation scheme, we chose to base the cartographic representation on familiar, well-labelled Ordnance Survey data.

6. The Public Participation Workshop

During the first period of system development, an opportunity for testing its use for public participation purposes emerged. This was carried out with a separate grant, which helped in establishing a network of experts with knowledge in planning, geomatics, GIS, public participation and inclusionary processes and environmental research. This element enabled the integrate the views and needs of this audience to the project scheme.

Public Participation GIS (PPGIS) is a current active research theme in GISc. The origins of this research are usually traced to from collaborative uses of GIS (Densham *et al.*, 1995) and GIS critique, epitomised by the publication of "Ground Truth" (Pickles, 1995). The main research themes emerge during NCGIA Initiative 19: GIS and Society (Harris and Weiner, 1996) where the concept itself was suggested and accepted (Schroeder, 1997)[3]. PPGIS has emerged as a test-bed for techniques, methodologies, ideas and discussion about the social implication of GIS technology. In recent years, this area has grown extensively and the project team consulted the material that emerged from this field during the design of the workshop. Though some previous activities focused on web-based PPGIS (Carver and Openshaw, 1995; Craig, 1998; Kingston *et al.*, 2000), the team involved in the Brownfield project preferred a more personal and contextual approach, which is common in collaborative planning research. There is some evidence for integrating similar methodologies in PPGIS (Al-Kodmany, 1998), but it was felt that the experience that was gathered in running various participative techniques, like in-depth group discussions or focus groups, could contribute to this field (see Burgess *et al.*, 1998a; Burgess *et al.*, 1988; Harrison *et al.*, 1998). Furthermore, the approaches that were used to design and run the workshop, were based on an inclusionary and participatory research agenda which relates to many areas in planning and governance (Burgess *et al.*, 1998b).

Based on these groundings, the workshop aimed to achieve two major goals. First, to enhance our understanding of users' need and requirement and secondly to explore the adaptability of participatory and inclusionary approaches that are more common in cultural geography research to PPGIS. In this paper, we will focus more on the requirements and needs, as the second aspect deserves a special and separate attention. As for the practical aspect of the workshop, we aimed to enable participants to learn something about GIS and its uses, while learning from them what they expect to find in such system. Fifteen people encompassing a wide range of computer skills were recruited from community and other voluntary groups based in Wandsworth.

The workshop took place at UCL and was held as a half-day session divided into four parts. The workshop started with introduction to the aims of the day and the technology.

---

[3] For a review of the origins and background, the interested reader is referred to (Schroeder, 1997) and (Chrisman, 1999a).

This introduction was as free from jargon as possible and gave an overview of the richness of information in the system while demonstrating basic GIS technology such as layering of information. Following this introduction, the participants divided into small groups so that they would all have an opportunity to use the GIS "hands-on". A 'GIS expert' (a person with experience with the software and the content of the system) and a facilitator (a person with expertise in group work) supported each group. Tape recorders were used to record the session and Lotus ScreenCam software was used to capture the operations of the computer system. This session lasted for over 90 minutes. Once this session ended and the participants gain familiarity with the system, its capabilities and content, we have braked for a well-earned lunch, in which all the facilitators and participants had an informal opportunity to raise issues and to share experiences.

The next part of the workshop was conducted in two groups, divided according to gender. Each group conducted a one-hour discussion on the views of the participants on the system and the systems use. The discussion tried to expand beyond the immediate experience of GIS use, as issues like public access to the system, accuracy and trust were raised. The reason for the gender divide was as a result of previous experience of the facilitators involved in previous group work. To conclude the day, a plenary session was held and provided more feedback through a debriefing questionnaire.

Participants felt that the system had some potential for use as a tool to provide them with a means of presenting their cases for local issues to planners and local authorities in a pro-active manner. They felt that the use of such a system might allow for a more informed debate between the public and local authority representatives on a more even footing. However, the public participation day did help to highlight a number of issues. Specifically, it proved that even novice users of GIS grasp the ideas of overlaying quickly and expected to see such analysis results. Furthermore, they quickly realise GIS ability to pursue and demonstrate 'what if?' scenarios of change

In terms of physical access to the information on the system, many of the participants felt that there were large sections of the population that might be excluded if the information was just provided over the Internet. One suggestion for a possible solution was that the Council should provide access to the system in local libraries. Participants were also concerned about the availability of software and the cost implications to a local group or individuals. Furthermore, some felt that many people would not have the necessary computer literacy or expertise to use the system and that if they had been left on their own, they would not have got very far. This would pose a real problem even if the system were available in libraries, as some people felt they would lack the confidence to ask for help. It is important to note that most of these aspects have been identified in the PPGIS literature (Kingston *et al.*, 2000) and that some have even provided an account for the use of GIS by local community group (Ghose, 1999).

As for the issue of information requirement, the workshop helped us in consolidating the data sets needs. The participants felt that the system would benefit from a range of additional data sets such as proposed parking schemes, traffic densities and flows, schools, cycle routes and landmarks such as local rivers. Furthermore, they felt that up-to-date information on population density; schools and other socio-economic variables would be helpful. Such information might enable pro-actively from the local community themselves. However, the network analysis, which showed up areas that were not well served by public transport, was seen as valuable. Participants wanted to be able to add their own information to the GIS so as to ensure that the planning process was not driven entirely by the concerns of 'experts'. Many were enthusiastic about the new possibilities for gaining new knowledge about their local area offered by the GIS. Noteworthy, some participants were unhappy with the level of data accuracy and reliability - especially the lack of local information (like local amenities or the type of shopping that each shopping area provides). On the organisational settings, participants were not convinced that updating of the system would be done regularly if it were the responsibility of the local authorities. Instead academic institutions were seen as reliable agents who could take responsibility for updating the GIS.

The PPGIS workshop identified some discrepancies in the system in relation to the availability of more detailed information on the surroundings of each brownfield site - it was noted that most group felt "disoriented" once the zoomed in to a specific area and the skeletal Meridian data set became too abstract. As a result, Aerial photographs were added around major brownfield sites.

7.   Exposing The System To Potential Users

After the workshop, the brownfield GIS was further developed and completed. Once the system was ready for presentation and discussion, a series of presentations were arranged. During these presentations, the system was presented to representatives of bodies and agencies with an interest in brownfield issues. Each presentation was followed by a 1-hour discussion on the merits and the discrepancies of the specific system and the use of GIS for this issue. The audience for the presentations included representatives from the Environmental Agency professional, both those involved in the day-by-day activities of the agency and the IT and GIS support personnel. Another presentation included representatives from local authorities and the Government Office for London.

A major issue that was raised by professionals who work daily with the planning system and with GIS was the issue of copyright and the cost of database creation and maintenance. These issues were seen as the major obstacles before the implementation of a full-scale brownfield GIS for London could be developed. The participants of the presentations agreed on the necessity to have such a database and that it would streamline brownfield development and improve the level of discussion. These presentations provided insights into the future uses and development of such a system by the most likely users.

8.   Discussion

The lessons from the brownfield GIS development process can be divided into three areas. First, we will comment on the nature of the development cycle and reflect on the state-of-the-art in GIS, GI and GISc. Second, we will comment on the contribution of GIS to the current debate about brownfield development. We will close with some comment and questions about current PPGIS research and the ways in which our project fits to the wider research agenda.

9.   Iterative Development

The system was developed using an iterative approach by using a series of seminars, demonstrations and a one-day public-participation workshop. This provided invaluable personal contact during the development process.

The use of commercial-off-the-shelf products helped in implementing the project in a very short time scale and with very limited resources (only one person was dedicated full-time to the project). The support and help from various stakeholders helped immensely in constructing a useful database. This iterative approach might seem an impediment to the development of the system, but instead it actually helped in discovering any missing data sets. For example, during a presentation to the workshop team, the need for postcode data sets emerged which was later integrated into the system.

This cycle is very similar to the concepts and ideas that are now commonplace in Rapid Application Development (RAD) (Reeve and Petch, 1999). In the brownfield GIS we have adopted RAD principles (rapid cycle of assembly, tests, exposure to the user community and so on) and given them equal importance in GIS database construction and development.

As the literature that we have reviewed in earlier sections demonstrates, we relied on techniques and approaches that were part of cutting-edge research project only 5 years ago. Indeed, the development in desktop GIS, the capabilities of the software that we have used answers most of the wish list that was described only few years ago for "True desktop GIS" (Elshaw-Thrall and Thrall, 1999). Furthermore, the proliferation of of-the-shelf, ready to use GI products (like Cities Revealed) makes the task of GIS database assembly much easier. True, our organisational settings meant that it was easier for us to obtain and use these GI data sets and others might find that the resources needed to purchase them are quite significant. This is an issue that often rises in the GI literature (Pipes and Maguire, 1997). However, as a proof of concept we have demonstrated that the infamous 80/20 equation between data acquisition cost and other cost of GIS project (Huxhold and Levinsohn, 1995), might be changing and soon may become irrelevant.

10. The Role Of GIS In The Brownfield Debate

The need to achieve a consensus among all relevant parties, including the public was stressed in Agenda 21, a global action programme for sustainable development (UN, 1992). The Rio Earth Summit (1992) recommended the active participation of citizens along with governments in the implementation of the Rio agreements. However, this is not easy as it implies the need to adopt new decision making methods that go beyond traditional consultancy frameworks to involve groups and individuals in a partnership approach. This approach emphasises the need to identify a range of issues and concerns and to resolve differences of opinion and conflicts between different interests, so that solutions are designed to meet various points of view (Environment Agency, 1998). The implementation of such an approach was tested as part of this project through the four Brownfield Seminars in February and the Public Participation workshop in June 1999.

The completed system provides a useful tool that demonstrates the advantages that GIS has over more traditional mapping methods; its ease of integrating diverse data sets, updating and Internet links to name a few. However, throughout its development many issues were raised especially relating to the points raised during the iterative development process.

By its very nature, a GIS is data-driven and its success is dependent on the availability of data. It is unfortunate that the NLUD is not more GI savvy and the integration of it with GIS will not be trouble free. However, when considering the urgency of the task to identify brownfield, it is believed that it can be integrated (even as point data) to a GIS scheme and therefore it will be possible to use it as part of analysis of the sort we described above. We have decided to obtain brownfield site information from the UDP's for each local authority, in spite of the lack of currency. In a way, we have demonstrated the potential of polygon-based NLUD.

We have selected an approach of rapid digitisation and integration of multiple data sets, knowing that the database accuracy and precision could be questionable. Some inaccuracy does occur in the digitising and geo-referencing process and as a result some of the data sets did not always correspond to each other perfectly. However, it is stressed that the system was not intended to be used by surveyors for pinpoint accuracy, but more as an information tool to try and encourage more public interest in the planning process and highlight the need for more sustainable development.

Finally, we should note that GIS holds the potential raised by those who have been exposed to the system (participants in the workshop, during the seminars and the presentation) that it should aim to predict the brownfield sites of tomorrow by considering sectors that are likely to produce new sites. Naturally, such information can be politically and commercially sensitive and their inclusion or exclusion is very subjective. However, such potential might be a challenging, yet interesting source for future research.

11. Public Participation GIS And Exploratory GIS

The approach that we have presented during the development of the brownfield GIS and in the course of this paper can be termed "discussion support tool". By this, we mean the use of GIS not just as a decision support tool, where alternatives are explored, quantified and compared using analytical models (Densham, 1991) but more akin to the Nicholas Negroponte concept of "tools to think with for the world at large" (Bennahum, 1995). The focus in this mode of GIS use is not necessary on its geodetic or analytic capabilities (although they do play a major role), but rather on the visual and contextual exploration of the problem situation and issues connected to it. This mode of use relies heavily on GIS capabilities to work "at the speed of discussion" (as one of the participants described it) and the ability of the expert user to understand, facilitate and perform the analysis on the spot using the full toolbox of GIS capabilities.

It is exactly at that point that our approach differs from the one offered by Web-based PPGIS ones (Carver *et al.*, 1998; Craig, 1998; Kingston *et al.*, 2000). Following previous discussion about expected functionality of GIS packages (see Elshaw-Thrall and Thrall, 1999) we argue that for a 'true' PPGIS a 'true' GIS is needed. Classic definitions of what GIS constitute (for example Maguire, 1991), the GIS principles offered by the International Association of Assessing Officials - IAAO (Huxhold and Levinsohn, 1995) all discuss the analytical capabilities as part and parcel of what GIS is. Earlier definition went as far as declaring that a specific functionality (polygon overlay) is needed as "proof" for GIS (Chrisman, 1999b). In the same paper Chrisman is offered a more open and inclusive definition of GIS, but also commented about definitions that are too inclusive (Rhind's):

"In his attempt to be universal, Rhind offers a definition that is so loose that the address book function of a hand-held pocket planner is indistinguishable from a full-function GIS workstation. ..." (Chrisman, 1999b, p. 181).

Similar emphasis on processing and interpretation of spatial data can be found in recent GIS textbooks (for example Heywood *et al.*, 1998). Therefore, some questions must be raised about the needed functionality of Web-based PPGIS. Current examples clearly lack these analytic capabilities (or carry it in a closed form as in Carver and Openshaw, 1995). Maybe the time has come to learn from the lessons of the development of general GIS and to open up the question of which functions are needed to make a 'true' PPGIS. This is not merely a question of definition. The importance of GIS is not in its capability to display interactive maps but in the ability to analyse spatial data. A better understanding of the analytical operations needed and the development of more accessible interfaces to perform these will provide some directions for future research.

Another important lesson about PPGIS is the opportunity that it opens for collaborative research among various expertise in Geography and related fields. One of the major reasons for the success of our workshop - at least as expressed by participants, was the integration of experience GISc researchers and researchers with a long track in participatory and inclusionary processes. We hope to report on this collaboration in more details soon[4].

From our experience we tend to agree with Harris and Wiener (Harris and Weiner, 1996) assertions as the fundamentals of successful PPGIS implementations:

- Agency driven, but not top-down nor privileged toward conventional expert knowledge
- Local knowledge is valuable and expert
- Broaden access base to spatial information technology and data
- Incorporate socially differentiated multiple realities of landscape
- Integrate GIS and multimedia
- Explore the potential for more democratic spatial decision making through greater community participation
- Assume that spatial decision making is conflict ridden and embedded in local politics

A valuable finding of our workshop is that many of those who are the potential users of PPGIS (like our workshop participants) identified some of these principles and expresses them without prior knowledge of the research literature.

12. Conclusions

Despite the issues related to accuracy and precision and the lack of current data, there are many advantages of the brownfield GIS over more traditional forms of data capture. By using the approach we described, we have managed to integrate different data sets so comparisons can quickly be made between brownfield sites and environmental data sets. Additional information such as ownership, size of the site and environmental and transport data can quickly and easily accessed each brownfield site in a web environment. It is assumed that the rollout of NLUD during 2000 will open-up possibilities to integrate direct access to its database, using a Web based interface similar to the one described above. The use of GIS can also enable the creation of a hierarchy of development suitability by comparing brownfield proximity to public amenities such as town centres and public transport.

Despite the drawbacks to the system, the finished product was well received by both. It is hoped that this project demonstrated what GIS can achieve for the brownfield debate and furthermore how important it is for the development process to be iterative. Planners

---

[4] For a preliminary report on the UCL Brownfield Research Network, see Aurigy *et al.* 1999.

need to investigate the spatial relationships between natural, physical and socio-economic variables to explore and evaluate different alternative planning scenarios. Therefore, the value of GIS for urban and environmental planners is its ability to integrate diverse data sets under a common spatial theme.

Acknowledgements

We would like to thank the Environment Agency and the London Ecology Unit for support and data, the Ordnance Survey for the use of their data in the development of the system, the London boroughs of Wandsworth, Sutton, Merton and Croydon and everyone that took part in the public participation meeting and seminars. Finally we would like to thank Dan Bloomfield for the original concept.

References

Al-Kodmany, K. (1998) GIS and the Artist: Shaping the Image of a Neighborhood in Participatory Environmental Design In *Empowerment, Marginalization, and Public Participation GIS*, Santa Barbara, CA,14-17th Oct.,1998 Available World Wide Web URL http://www.ncgia.ucsb.edu/varenius/ppgis/papers/al-kodmany.html (Accessed 3rd May 1999).

Armstrong, M. P., Densham, P. J., Lolonis, P. and Rushton, G. (1992) Cartographic Displays to Support Locational Decision Making, *Cartography and Geographic Information Systems,* **19**(3), pp. 154-164.

Aurigi, A., Batty, S., Bloomfield, D., Boott, R., Clark, J., Haklay, M., Harrison, C., Heppell, K., Moreley, J. and Thornton, C., 1999, *UCL Brownfield Research Network*, University College London, London.

Batty, M. (1993) Using Geographic Information Systems in Urban Policy and Policy-Making, In *Geographic Information Systems, Spatial Modelling and Policy Evaluation,*(Eds. Fischer, M. M. and Nijkamp, P.), Springer-Verlag,Berlin, pp. 51-72.

Bennahum, D. S. (1995), *Meme 1.07*, Available E-mail newslaetter, URL: http://www.memex.org/meme1-07.html (Accessed 26th Aug 1999)

Bloomfield, D. (forthcoming) Towards an Evaluative Framework for Public Policy Discourses: London's Brownfields and a New Governance Relation, PhD thesis, University of London, London.

Breheny, M and Hall, P (1996) *The people – Where will they go?* TCPA, London

Brown, I. M. (1999) Developing a Virtual Reality User Interface (VRUI) for Geographic Information Retrieval on the Internet, *Transactions in GIS,* **3**(3), pp. 207-220.

Burgess, J., Clark, J. and Harrison, C. M. (1998a) Respondents' Evaluations of a CV Survey: A Case Study Based on an Economic Valuation of the Wildlife Enhancement Scheme, Pevensey Levels in East Sussex, *Area,* **30**(1), pp. 19-27.

Burgess, J., Harrison, C. M. and Filius, P. (1998b) Environmental Communication and the Cultural Politics of Environmental Citizenship, *Environment and Planning A,* **30**(8), pp. 1445-1460.

Burgess, J., Harrison, C. M. and Limb, M. (1988) People, Parks and the Urban Green: A Study of Popular Meanings and Values for Open Spaces in the City, *Urban Studies,* **25**(6), pp. 455-473.

Burrough, P. A. (1986) *Principles of Geographical Information Systems for Land Resources Assessment,* Clarendon Press, Oxford.

Carver, S., Kingston, R. and Turton, I. (1998), *Accessing GIS over the Web: An Aid to Public Participation in Environmental Decision Making*, Available World Wide Web, URL: http://www.ccg.leeds.ac.uk/vdmisp/publications/paper1.html (Accessed 11th Dec 1998)

Carver, S. and Openshaw, S. (1995) Using GIS to Explore the Technical and Social Aspects of Site Selection In *Conference on the Geological Disposal of Radioactive Wastes*, Royal Lancaster Hotel, London, March 1995.

Chrisman, N. (1999a) Full Circle: More Than Just Social Implications of GIS In *GISOC99*Ed,^(Eds, ence) Full Circle: More Than Just Social Implications of GIS University of Minnesota,20-22 June,1999a Available URL: http://www.geog.umn.edu/gisoc99/chrisman.htm (Accessed 5th Jul 1999).

Chrisman, N. R. (1999b) What Does 'GIS' Mean?, *Transactions in GIS,* 3(2), pp. 175-186.

Craglia, M. and Raper, J. (1995) GIS and Multimedia, *Environment and Planning B,* 22(6), pp. 634-636.

Craig, W. J. (1998) The Internet Aids Community Participation in the Planning Process In *Groupware for Urban Planning* Lion, France,5th Feb 1998.

Densham, P. J. (1991) Spatial Decision Support Systems, In *Geographical Information Systems: Principles and Applications,*(Eds. Maguire, D. J., Goodchild, M. F. and Rhind, D.), Longman Scientific and Technical, Harlow, England, pp. 403-412.

Densham, P. J., Armstrong, M. P. and Kemp, K. K. (1995), *Collaborative Spatial Decision-Making: Scientific Report for the Specialist Meeting*, National Center for Geographic Information and Analysis, Santa Barbara, CA.

Doyle, S., Dodge, M. and Smith, A. (1998) The Potential of Web-Based Mapping and Virtual Reality Technologies for Modeling Urban Environments, *Computers, Environment and Urban Systems,* 22(2), pp. 137-155.

Elshaw-Thrall, S. and Thrall, G. I. (1999) Desktop GIS Software, In *Geographical Information Systems,*(Eds. Longley, P., Goodchild Michael, F., Maguire, D. J. and Rhind, D.), John Wiley & Sons Inc.,New York, pp. 331-345.

Environment Agency, (1998) *SD12 Consensus Building for Sustainable Development*, Sustainable Development publication series.

Fonseca, A., Gouveia, C., Camara, A. and Silva, J. P. (1995) Environmental Impact Assessment with Multimedia Spatial Information Systems, *Environment and Planning B,* 22(6), pp. 637-648.

Ghose, R. (1999) Use of Information Technology for Community Empowerment: Transforming Geographical Information Systems into Community Information Systems In *GISOC99* University of Minnesota, 20-22 June,1999 Available World Wide Web URL: http://www.geog.umn.edu/gisoc99/ (Accessed 5th Jul 1999).

Goodchild, M. F., Parks, B. O. and Steyaert, L. T. (Eds.) (1993) *Environmental Modeling with GIS*, Oxford University Press, New York.

Harris, T. M. and Weiner, D. (1996), *GIS and Society: The Social Implications of How People, Space and Environment Are Represented in GIS*, National Center for Geographic Information and Analysis, Santa Barbara,CA.

Harrison, C. M., Burgess, J. and Clark, J. (1998) Discounted Knowledges: Farmers' and Residents' Understandings of Nature Conservation Goals and Policies, *Journal o Environmental Management,* 54(4), pp. 305-320.

Healey, P. (1997) *Collaborative Planning : Shaping Places in Fragmented Societies,* UBC Press, Vancouver.

Healey, P. (1998) Collaborative Planning in a Stakeholder Society, *Town Planning Review,* **69**(1), pp. 1-21.

Heuvelink, G. B. M. (1998) *Error Propagation in Environmental Modelling with GIS,* Taylor & Francis, London.

Heywood, I., Cornelius, S. and Carver, S. (1998) *An Introduction to Geographical Information Systems,* Longman, Harlow.

Huxhold, W. E. and Levinsohn, A. G. (1995) *Managing Geographic Information System Projects,* Oxford University Press, New York.

Kingston, R., Carver, S., Evans, A. and Turton, I. (2000) Web-Based Public Participation Geographical Information Systems: An Aid to Local Environmental Decision-Making, *Computers,Environment and Urban Systems,* **24**(2), pp. 109-125.

Landauer, T. K. (1995) *The Trouble with Computers: Usefulness, Usability, and Productivity,* MIT Press, Cambridge, Mass.

Laurini, R. and Thompson, D. (1992) *Fundamentals of Spatial Information Systems,* Academic Press, London.

Maguire, D. J. (1991) An Overview of Definition of GIS, In *Geographical Information Systems: Principles and Applications,*(Eds. Maguire, D. J., Goodchild, M. F. and Rhind, D.), Longman Scientific and Technical,Harlow, England, pp. 9-20.

Martin, D. (1991) *Geographic Information Systems and Their Socioeconomic Applications,* Routledge, London, Routledge. 1991.

McHarg, I. L. (1969) *Design with Nature,* Published for the American Museum of Natural History [by] the Natural History Press, Garden City, N.Y.

Medyckyj-Scott, D. and Hearnshaw, H. M. (1993) *Human Factors in Geographical Information Systems,* Belhaven, London.

NLUD - National Land Use Database, 2000, Previously Developed Land (PDL) - Data Specification, Available World Wide Web http://www.nlud.org.uk (Accessed 1st Jul 2000).

Nyerges, T. L., Mark, D. M., Laurini, R. and Egenhofer, M. J. (Eds.) (1995) *Cognitive Aspects of Human-Computer Interaction for Geographic Information Systems,* Kluwer Academic Publishers, Dordrecht, The Netherlands.

Pickles, J. (1995) *Ground Truth: the Social Implications of Geographic Information Systems,* Guilford Press, New York.

Pipes, S. and Maguire, F. (1997) Behind the Green Door In *Mapping Awareness,* **11**(10), pp. 28-29.

Preece, J. (1995) *Human-Computer Interaction,* Addison-Wesley Pub. Co, Wokingham, England.

Raper, J. F., Rhind David, W. and Shepherd, J. (1992) *Postcodes : The New Geography,* Longman Scientific & Technical, Harlow, Essex.

Rasmussen, J. (1995) Geographic Information Systems, Work Analysis, and System Design, In *Cognitive Aspects of Human-Computer Interaction for Geographic Information Systems,*(Eds. Nyerges, T. L., Mark, D. M., Laurini, R. and Egenhofer, M. J.), Kluwer Academic Publishers,Dordrecht, The Netherlands, pp. 373-391.

Reeve, D. E. and Petch, J. R. (1999) *GIS, Organisations and People : A Socio-Technical Approach,* Taylor & Francis, London.

Schroeder, P. (1997) GIS in Public Participation Settings In *UCGIS 1997 Annual Assembly and Summer Retreat* Bar Harbor, Maine,15th-21st June,1997 Available World Wide Web URLhttp://www.spatial.maine.edu/ucgis/testproc/schroeder/UCGISDFT.HTM (Accessed 16th Aug 1999).

Shiffer, M. J. (1995) Environmental Review with Hypermedia Systems, *Environment and Planning B,* **22**(3), pp. 359-372.

Tomlinson, R. F. (Ed.) (1970) *Environment Information Systems,* International Geographical Union, Ottawa, Canada.

UN (1992), *Agenda 21,* United Nations, Rio de Janeiro.

Urban Task Force (1999*) Toward Urban Renaissance*, DETR, London

16

Evaluation Methods to Support the Comparison of Maps for Environmental Decision Making

Marjan van Herwijnen, Ron Janssen

1. Introduction

In environmental management the performance of policy alternatives is often represented in maps. These maps show the spatial distribution of the impact of policy alternatives. The decision maker is asked to use these maps to compare alternatives and ultimately to select the one preferred. Since most people use reference maps, such as road maps, maps are familiar to people and are expected to represent reality. Decision makers are therefore happy to use maps to support their decisions (see for example Kohsiek et al. 1991).

In practice effective use of maps is a difficult task for many people and gets more difficult if the information density of the maps increases and the direct link with reality decreases (Muehrcke and Muehrcke, 1992). It is known that people are able to process no more than seven stimuli at the time (Miller, 1956), and that the interpretation of images is strongly dependent on map design (Bertin, 1981; Monmonier, 1991; MacEachren, 1994; Kraak and Ormeling, 1996). Comparison of map representations of alternatives is therefore an extremely difficult task which gets exponentially more difficult if more than two maps need to be compared at the same time.

This paper first explains some theoretical concepts of spatial multi-criteria analysis. A nature management problem in the Green Heart of the Netherlands is used to show the potential of spatial decision support (see also Herwijnen et al., 1997 and Herwijnen, 1999). Five examples of policy questions illustrate the use of spatial evaluation methods.

2. Multiple Objectives In Space

Problems with multiple objectives can be evaluated using multi-criteria analysis (MCA). MCA can be used to compare alternative policy options. It provides a systematic, transparent approach that increases objectivity and generates results that can be reproduced (Bonte *et al.* 1998). The impact on the objectives is measured using criteria.

The performance of alternative policy options can be presented as a table of values (an effects table). Figure 1a is a schematic representation of an effects table and shows the effects of three alternative policy options (a_1, a_2 and a_3) for four different criteria (c_1, c_2, c_3 and c_4). The effects of each policy alternative for each criterion is measured by a single value, represented here by a dot.

Environmental management makes increasing use of spatial simulation models to predict the effects of alternative options. These models result in a prediction of the spatial distribution of the impacts. This implies that the effect of an alternative option for a criterion is not just one value, as visualised in Figure 1a, but a matrix of values. These values can be visualised as a map. Consequently, the non-spatial effects table, which is a table of values (Figure 1a), is transformed into a spatial effects table, which is a table of maps (Figure 1b).

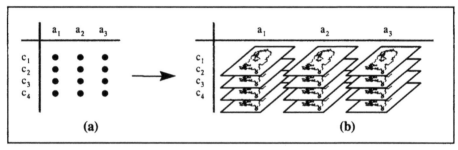

Figure 1. *Transformation of a non-spatial effects table (a) into a spatial effects table (b).*

The spatial decision problems handled in this paper must not be confused with another type of spatial decision problem: the ranking of alternative locations within one map. The question in facility location problems is where to locate certain activities such as a public facility, a residential area, an industrial site, etc. In a GIS environment, these problems are usually processed using weighted overlay or other standard multi-criteria methods resulting in a ranking of the spatial units used. Only if the surroundings of a location influence the value of that location, can the problem be expressed in the type of spatial decision problem handled in this paper. Examples of the evaluation of ranking alternative locations within one map can be found in Ghosh and Rushton (1987), Carver (1991), Eastman et al. (1995), Pereira and Duckstein (1993), Davidson et al. (1994), and Eastman (1997).

3. Spatial Multi-Criteria Analysis

Conventional, multi-criteria analysis starts with an effects table like the one in Figure 1a. The criteria are used to measure the effects of the alternatives. This is not always done using the same measurement scale. Therefore, the first step of most multi-criteria methods is to make the effects of the alternatives for the specified criteria comparable. This can be done using various means of standardisation. The next step is to determine the relative importance of the criteria. Finally, a ranking of the alternatives can be generated using a decision rule. These rules depend on the multi-criteria method used.

In environmental management the performance of a policy alternative is often presented in map form instead of a set of values. This increases the amount of information to be evaluated enormously. The decision maker is asked to use these maps to compare alternatives and ultimately to select the one preferred. A major part of the information included in these maps is usually generated by simulation models. This part of the information only represents the real world under the specific assumptions used by the models. As a consequence, the direct link between the (spatial) performance of the alternatives and reality is low.

Comparison of map representations of alternatives is therefore an extremely difficult task which gets exponentially more difficult if more than two maps are to be compared at the same time. Spatial evaluation methods, implemented in a GIS, are designed to support this task. They help the decision maker to structure and simplify the spatial performance of the alternatives. The starting point is the 4-dimensional effects table (Figure 1b). This table contains alternatives (a), criterion scores (c) and the geographical location of the scores (x,y). The objective of the evaluation can be: the selection of a best alternative; the selection of a number of acceptable alternatives; or, a complete ranking of the alternatives.

Figure 2 shows three paths which can be taken to aggregate the 4-dimensional effects table (upper-left corner) into a ranking of the alternatives (lower-right corner). Following Path 3, all available information is offered to the decision maker and the decision maker alone has to process this information into a ranking. In Path 1 and Path 2, the aggregation of the spatial effects table is carried out in two sub-steps: spatial aggregation (SA) and multi-criteria analysis (MCA). The two paths differ according to the order of the two sub-steps. Path 1 first aggregates space from the 4-dimensional table resulting in a standard effects table (upper-right corner). In the second sub-step, this table is processed using standard MCA resulting in a ranking of the alternatives. Path 2 first combines the multiple criteria into overall spatial data sets representing the overall performance for each alternative (lower-left corner). Then, in the second sub-step, space is aggregated from the overall spatial data sets resulting in a ranking of the alternatives.

Only a few methods are available to support Path 3. For instance, one can think of methods to make the information more accessible by means of graphical or cartographical tools (see, for example, Tufte, 1985, 1990; Kraak and Ormeling, 1996). Methods for the spatial aggregation in Path 1 of the spatial effects table (SA) can be derived from the large variety of spatial analysis methods described in the literature (Bartlett, 1975; Burrough and McDonell, 1998; Campbell, 1991; Cressie, 1993; Haining, 1994; Hearnshaw and Unwin, 1994; Openshaw, 1991; Unwin, 1981; Upton and Fingleton, 1985). The next sub-step, evaluating the non-spatial effects table into a ranking (MCA), can be done using standard multi-criteria methods (Beinat, 1997; Janssen, 1992; Janssen and van Herwijnen, 1994; Keeney, 1992). With small adjustments, some of these methods can also be used for the first sub-step in Path 2: combining the criteria of the spatial effects table into overall spatial data sets representing the overall performance of the alternatives (Beinat and Janssen, 1996; Eastman, 1997; Eastman et al., 1995; Tkach and Simonovic, 1997). Examples of combining both steps in each of the paths are rare. Herwijnen et al. (1993a) is an example of both steps in Path 2 and Janssen and Rietveld (1990) is an example of both steps in Path 1.

Path 3 is most common in practice. The spatial effects table is presented to the policy maker as a set of maps without further support. If the table contains many and complex information this can lead to wrong conclusions. In this case path 1 or path 2 should be selected. The choice between path 1 and 2 depends on the spatial characteristics of the decision problem. The advantage of path 1 is that the spatial pattern of each criterion separately can be included in the spatial aggregation. Examples are the calculation of connected natural areas or the size of the area with a value above a certain threshold for a single criterion. Path 2 produces a map that shows the total value of all criterion scores

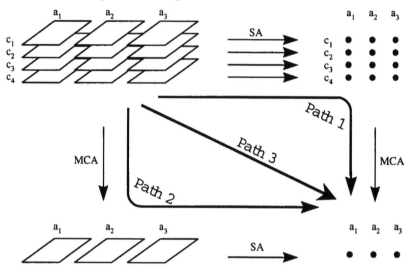

Figure 2. *Three paths to rank alternatives with spatial impacts.*

combined. This is preferred if the spatial pattern of the combined criterion scores is relevant. Path 2 also makes it possible to concentrate on specific areas. In conclusion, both paths have their limitations and selection of the best path is dependent on the decision problem. Problems with an interest in zoning are preferably evaluated using Path 2. Path 1 is preferred if the interest is not so much on the zones themselves but on the spatial patterns of these zones. More detailed information about the differences and similarities among these paths can be found in Herwijnen and Rietveld (1999). A mathematical description of these paths can be found in Herwijnen (1999).

In most cases of environmental management, the size of a problem makes the use of GIS indispensable. Only then can the information be processed so that it is useful for a decision maker. This is supported by Crossland et al. (1995), who investigated whether the addition of GIS technology to a decision-making environment affects the performance of the individual decision maker when the decision task involves spatially referenced information. In a laboratory experiment these authors manipulated the availability of SDSS (available or not) and the problem complexity (more or less complex). The problems used were objective and had a predetermined correct ranking as the solution. They measured the time needed to solve the problem and the accuracy of the retrieved ranking using 142 people. They concluded that GIS users need less time to make the decision than non-GIS users and that GIS users have an increased accuracy compared with non-GIS users.

4. An Example Of The Use Of Spatial Multi-Criteria Analysis: Nature Management In The Netherlands

The use of spatial multi-criteria analysis is illustrated by means of a management problem in the Green Heart of the Netherlands (see also Herwijnen *et al.* 1997). The evaluation in this case study deals with the area as a whole. The spatial pattern of the impacts plays an important role. Therefore, Path 1 is selected to evaluate the alternatives. The evaluation methods developed are integrated in the Decision Support System Nature Management. This DSS is developed by the RIVM (Latour et al., 1997).

The Green Heart is an area of relatively open space in the centre of the four cities of Amsterdam, Rotterdam, The Hague and Utrecht (see Figure 4). The Green Heart has an agricultural landscape but there are also areas which have important natural qualities. At present such natural areas are very fragmented and their quality is declining. This is leading to a decline in species diversity. It is essential, therefore, to create larger *connected* natural areas offering opportunities for rare species to develop, thus ensuring sustainable natural qualities.

Figure 3 presents the structure of the evaluation module in the DSS Nature Management. The starting point of this DSS is *policy alternatives* containing measures to reduce emissions of SO_2, NOx and NH_3. The alternatives differ in the intensity and spatial pattern of these measures. A spatial *simulation model* is used to calculate the probability of occurrence of various plant species and to predict the number of plant species that occur per type of ecosystem. The *model result* for every alternative is the potential quality of all ecosystems with a spatial resolution of 1 km^2 (Kross et al., 1995; Latour et al., 1997).

The model results are confronted by the spatial *policy objective* presented in Figure 4 (Latour et al., 1998). This policy objective establishes the size and spatial pattern of the desired ecosystems for the Green Heart. The policy plan is only defined for rural areas and so non-rural areas are white in Figure 4. The spatial policy plan is defined for 813 km^2 whereas the total area for the Green Heart is 1862 km^2.

Figure 4 shows that a few ecosystems have been planned for large connected areas in the North-East and Centre of the Green Heart but that most ecosystems have been planned as small isolated units. Some of these are allocated only a few square kilometers (Latour et al., 1997).

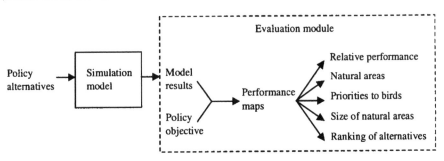

Figure 3. *The evaluation module in the DSS Nature Management.*

Comparison of the spatial objective with the model results leads to a number of *performance maps*. These maps contain the extent to which the different alternatives realise the planned ecosystems. The evaluation module is designed to support decision makers to compare these performance maps. The module focuses on a separate comparison for each evaluation objective, and on a total comparison. The following five examples of policy questions illustrate the use of the spatial evaluation methods used:

1. What is the relative performance of the alternatives?
2. How does the relative performance change if the comparison is linked to quality thresholds?

3. How does the relative performance change if some types of nature are considered more important than others?
4. Which of the alternatives generates the largest connected natural areas?
5. What is the best alternative?

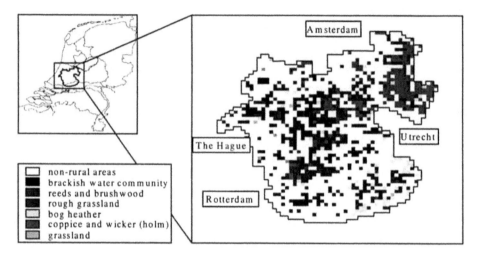

Figure 4. *A spatial policy objective for the Green Heart of the Netherlands.*

5. Relative Performance

The performance maps shown in Figure 5 represent the potential nature quality for each alternative. The values linked to each grid cell in these maps originally range from 0 (none of the plant species in the specified ecosystem can survive) to 100 (all can survive). These maps could be represented on a continuous scale as proposed by Tobler (1973). However, research has shown that the readability of the map is reduced by too many classes which make it difficult, at a single glance, to get an overview and understanding of the theme being mapped (Kraak and Ormeling 1996). The maps represented in Figure 5 are therefore classified in three classes. Differences among alternatives can be obscured by the use inadequate class limits. Limiting the number of classes and at the same time changing class limits may provide a better image of the differences among the alternatives. A difference index is used to determine the optimal class limits.

Each map is evaluated using an evaluation index. The index value of the map representing the policy objective is set at 100 for all evaluation criteria. Definition of the index depends on the aspect represented in the map. The performance index included in Figure 5 represents the average performance of the alternatives. This index is independent of the classification used.

6. Quality Thresholds

How much nature of a certain minimum quality results from the alternatives? The answer to this question depends on the quality thresholds used. Is an ecosystem considered to exist if 80% of its plant species are found, or is 50% sufficient? Applying the quality thresholds on a performance map results in a total area map. By changing the quality thresholds the influence on the total area map can be examined.

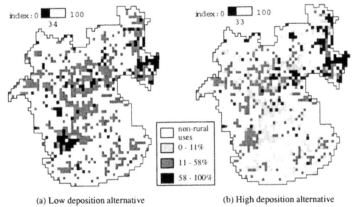

(a) Low deposition alternative   (b) High deposition alternative

Figure 5. *Predicted performance of the alternatives presented in three unequal classes*

Thresholds are also used to calculate indices for variation and naturalness. Variation is defined as the number of ecosystems specified in the policy objective that can exist in each alternative. An ecosystem exists if at least 3 km^2 of an ecosystem exceed the quality threshold. To calculate naturalness the area with specific natural qualities is calculated.

7. Priorities

So far it is assumed that all km^2 and all ecosystems are equally important. However, decision makers can have a preference for specific rare or characteristic ecosystems or for protection of certain areas such as nature reserves. In this example the relative weight of each ecosystem is derived from the number of bird species that use this ecosystem as preferred habitat for at least part of their life cycle. The weights are standardised in such a way that the average weight of the total policy area is equal to 1. Linking these weights to the spatial distribution of the ecosystems specified in the spatial policy objective results in the priority map shown in Figure 6.

Changes in the relative weights do not influence the performance maps (Figure 5). However changing the weights does influence the value of the evaluation indices linked to these maps. As was shown in Figure 5 the evaluation index is almost the same if all ecosystems are considered equally important (Low deposition: 45; High deposition: 44). If the above weights for each ecosystem are used these values change to 39 for the low deposition alternative and 33 for the high deposition alternative. The low deposition alternative is clearly the preferred alternative using these weights. Since the average weight for the whole policy area equals one, this implies that the higher weighted ecosystems perform badly in the high deposition alternative.

8. Large Natural Areas

At present nature is very fragmented and its quality is declining. This is leading to a decline in species diversity. It is essential to create larger connected areas to offer opportunities for rare species and to ensure sustainable natural qualities. It is therefore important to compare the size of connected areas in the alternatives.

Connected areas are calculated using King's neighbourhood. This implies that each grid cell can be connected to a maximum of eight other grid cells. Figure 7 shows the size

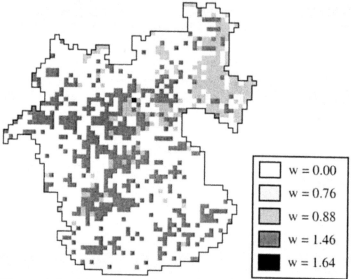

☐	w = 0.00
☐	w = 0.76
▨	w = 0.88
▨	w = 1.46
■	w = 1.64

Figure 6. *Priority map linked to number of bird species in each ecosystem*

of connected natural areas. In this figure no distinction is made among the ecosystems. This is a valid assumption for species that can survive in all different ecosystems. However if species are dependent on one ecosystem, connected areas can be calculated for each ecosystem separately. It is also possible to generate the largest connected area for each ecosystem. This is relevant if thresholds are specified for target species. The index is calculated as the total natural area divided by the square root of the number of separate areas to reflect both size and fragmentation.

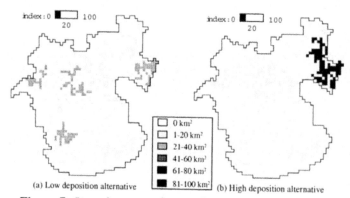

(a) Low deposition alternative

☐	0 km²
☐	1-20 km²
▨	21-40 km²
▨	41-60 km²
■	61-80 km²
■	81-100 km²

(b) High deposition alternative

Figure 7. *Size of connected natural areas*

9. Ranking

Each map is evaluated using an evaluation index. This index is presented with the map. Definition of the index is dependent on the aspect represented in the map. The index value of the map representing the policy objective is equal to 100 in all cases. Calculation of an evaluation index involves aggregation over space. The evaluation index can therefore be described as a non-spatial summary of the underlying map. Evaluation indices are included in an evaluation table. The evaluation table can therefore be described as a non-spatial summary of the decision problem and can be evaluated using a standard multicriteria approach (Janssen and van Herwijnen 1994).

Four classes of evaluation criteria are used: size, quality, spatial pattern and costs. The evaluation criteria are included in the evaluation table Table 1). The evaluation table shows the value of each alternative on a 0-100 scale. The last two columns show the value of 100 linked to the policy objective and the value before standardisation that corresponds to 100. For example the policy objective specifies five ecosystems. The maximum value of variation therefore corresponds to a total number of five ecosystems. The score of 80 for the Low deposition alternative therefore corresponds to the occurrence of four of the specified ecosystems. The first three of the above classes of evaluation criteria reflect the effectiveness of the alternatives to generate natural qualities. The last includes the cost of implementation of these alternatives. The numbers presented are tentative and represent a very rough estimate of the cost of instruments to reduce NH_3 emissions in and around the region (Herwijnen *et al.* 1997).

	Low dep.	High dep.	Max.	Max. represents
Size				
Total area	45	44	100	813 km^2
- priority to conserve areas	46	44	100	813 weighed km^2
- priority for bird species	39	33	100	813 weighed km^2
Quality				
Performance	34	33	100	all spec. plant species
Variation	80	60	100	5 different ecosystems
Naturalness	47	35	100	513 km^2 semi-natural area
Spatial pattern				
connected areas/nature type	55	53	100	15.1 km^2
connected natural areas	37	36	100	102 km^2
largest connected areas	46	52	100	largest area: 49.8 km^2
Cost				
Total costs	1.85	0.59	5	5 million guilders/year
Effectiveness	46.7	42.5	100	
Cost-effectiveness	25.2	72.1	∞	

Table 1. *Evaluation table*

All scores representing effectiveness are measured on the same scale (0-100) and can therefore be combined into an overall effectiveness index. All evaluation criteria make use of the information in the performance maps (Figure 5). This information is combined with other information such as thresholds and weights. Evaluation criteria also use different attributes of the same underlying data. Therefore the scores are, to a certain degree, interdependent especially within each category. The contribution that each of the criteria makes to the overall index is reflected in weights. These weights do not represent trade-offs, but represent the influence of the criteria in the overall objective. An effectiveness score for each alternative is calculated as the weighted sum of the scores of the evaluation criteria.

The importance of the individual criteria for the overall score is dependent on the function of the region. Because the Green Heart plays an important role as a buffer separating large urban areas, criteria linked to size are considered important. In this example these criteria have been given twice the weight of the quality and pattern-related criteria. These weights result in the effectiveness score presented in Table 1. The effectiveness scores show that the Low deposition alternative is more effective than the High deposition alternative. It is also shown that both alternatives are far removed from the policy objective. Only if all results are attributable to the alternatives can cost-effectiveness be calculated as the ratio of effectiveness and costs. In practice adjustments will have to be made for autonomous developments. In this case the High deposition alternative is the most cost-effective while achieving the policy objective is the least cost effective.

10. Conclusions

The results of model calculations were presented in performance maps. These maps represent the relative quality of the alternatives and are the input to the evaluation. From the performance maps the differences between the alternatives were not obvious. Only by looking carefully do differences become clear. It was shown that the alternatives differ if priority is given to bird species and also that the alternatives differ substantially in spatial pattern. Each map was summarised using an evaluation index. These indices were included in an evaluation table. From this table it could be concluded that the differences in effectiveness between the alternatives were small. However, important differences between the alternatives were lost in aggregation. Differences that could be observed from the maps were not always reflected in the indices. Therefore maps and indices should be used in combination.

The resulting performance maps in this case study were evaluated following Path 1. Insights were provided, not only by the ranking of alternatives from Path 1, but also from the intermediate criteria maps. These maps showed information about differences in location of the alternatives; this information was lost in the final ranking of the alternatives. Thus, the whole process from 4-dimensional evaluation table to a final ranking of alternatives helps to understand and compare the resulting performance maps.

The results of this study offer many opportunities for further research. In the current study the spatial pattern of the performance maps and the policy objective is identical. Objective and performance only differ in level of achievement. As a next step performance maps with spatial patterns that differ from the policy objective could be evaluated using distance metrics. Specification of the policy objective is a complex task. Heuristics could be developed to support the decision maker in translating non-spatial objectives into an objective map. A clear reference point for the required natural qualities offers possibilities for increasing the cost-effectiveness of environmental policies. Optimisation procedures could be developed to generate optimal combinations in intensity, spatial pattern and time of emission reduction and management instruments.

Acknowledgement

The research presented in this article was commissioned by the National Institute of Public Health and Environmental Protection (RIVM), Bilthoven and was included in the MAP research project 'Environmental quality of natural areas'. The authors would like to thank Arthur van Beurden, Rob van de Velde and Joris Latour (RIVM) for their contributions to the project. The authors also thank Jos Boelens who created the computer program, Xander Olsthoorn who provided estimates of costs, and the anonymous reviewer who made valuable remarks about this paper.

References

Bartlett, M.S. (1975). *The Statistical Analysis of Spatial Pattern*, Chapman and Hall, London.

Beinat, E. and R. Janssen (1996). Decision support and spatial analysis for risk assessment of new pesticides. In: R. Mcmillan and H.F.L. Ottens (eds.), *Geographical information: from research to application through co-operation*, IOS Press, Amsterdam, pp. 757-766.

Beinat, E. (1997). *Value functions for environmental management*, Kluwer Academic Publishers, Dordrecht.

Bertin, J. (1981) *Graphics and graphic information processing*, Walter de Gruyter, Berlin.

Bonte, R.J., R. Janssen, R.H.J. Mooren, J.T.d. Smidt and J.J.v.d. Burg (1998) Multicriteria analysis: making subjectivity explicit, in: Commissie voor de milieueffectrapportage, (Ed.), *New experiences on environmental impact assessment in the Netherlands: process, methodology, case studies*, Utrecht, Commissie voor de milieueffectrapportage.

Burrough, P.A. and R.A. McDonnell (1998). *Principles of Geographical Information Systems*, Oxford University Press, Oxford.

Campbell, J. (1991). *Map Use and Analysis*, Wm.C.Brown Publishers, Dubuque.

Carver, S.J. (1991) Integrating multi-criteria evaluation with geographical information systems. *International Journal for Geographical Information Systems*, Vol. 5, No. 3..

Cressie, N.A.C. (1993). *Statistics for spatial data*, John Wiley & Sons, Inc., New York.

Crossland, M.D., B.E. Wynne and W.C. Perkins (1995) Spatial decision support systems: An overview of technology and a test of efficency, in *Decision Support Systems*, Vol. **14**.

Davidson, D.A., S.P. Theocharopoulos and R.J. Bloksma (1994) A land evaluation project in Greece using GIS and based on Boolean and fuzzy set methodologies, in *International Journal for Geographical Information Systems*, Vol. **8**, No. 4.

Eastman, J.R., W. Jin, P.A.K. Kyem and J. Toledano (1995) Raster procedures for multi-criteria/multi-objective decisions, *in Photogrammetric Engineering and Remote Sensing* , Vol. **61**, No. 5.

Eastman, J.R. (1997) *IDRISI for Windows: User's Guide, Version 2*, MA: Clark University, Graduate School of Geography, Worchester.

Ghosh, A. and G. Rushton (1987) *Spatial Analysis and location-allocation models*, Van Nostrand Reinhold Company, New York.

Goodchild, M.F. (1993) From Modeling to Policy, in: M.F. Goodchild, B.O. Parks and L.T. Steyaert (Eds.), *Environmental modeling with GIS*, Oxford University Press, New York.

Haining, R. (1994). Designing spatial data analysis modules for geographical information systems. In: S. Fotheringham and P. Rogerson (eds.), *Spatial Analysis and GIS*, Taylor & Francis, London, pp. 45-64.

Hearnshaw, H.M. and D.J. Unwin (eds.) (1994). *Visualization in geographical information systems;* Introduction: the process, John Wiley & Sons, West Sussex.

Herwijnen, M. van, R. Janssen and P. Nijkamp (1993). A multi-criteria decision support model and geographic information system for sustainable development planning of the Greek islands. *Project Appraisal*, Vol. March, No. 1993, pp. 9-22.

Herwijnen, M.van, R. Janssen, A.A. Olsthoorn and J. Boelens (1997*) Ontwikkeling van ruimtelijke evaluatiemethoden voor gebiedsgericht milieu- en natuurbeleid*, Instituut voor Milieuvraagstukken, Amsterdam.

Herwijnen, M. van and P. Rietveld (1999). Spatial Dimensions in Multicriteria Analysis. In: *Spatial Multicriteria Decision Making and Analysis; A Geographic Information Sciences Approach.* J.C. Thill (ed.), Ashgate, Brookfield, pp. 77-102.

Herwijnen, M. van (1999). *Spatial Decision Support for Environmental Management.* Vrije Universiteit, Amsterdam.

Janssen, R. and P. Rietveld (1990). Multicriteria analysis and GIS; an application to agricultural land use in the Netherlands. In: H.J. Scholten and J.C.H. Stillwell (eds.), *Geographical Information Systems and Urban and Regional Planning*, Kluwer Academic Publishers, Dordrecht, pp. 129-139.

Janssen, R. (1992). *Multiobjective decision support for environmental management*, Kluwer Academic publishers, Dordrecht.

Janssen, R. and M. van Herwijnen (1994*). DEFINITE A system to support decisions on a finite set of alternatives* (Software package and user manual), Kluwer Academic Publishers, Dordrecht.

Keeney, R.L. (1992). *Value-Focused Think*ing, Harvard University Press, Cambridge.

Kohsiek, L., F.v.d. Ven, G. Beugelink and N. Pellenbarg (1991) *Sustainable Use of Groundwater; Problems and threats in the European Communities*, RIVM/RIZA, Bilthoven.

Kraak, M.J. and F.J. Ormeling (1996). *Cartography: Visualization of Spatial Data*, Addison Wesley Longman, Harlow.

Kross, J., G.J. Reinds, W.d. Vries, J.B. Latour and M.J.S. Bollen (1995). *Modelling of soil acidity and nitrogen availability in natural ecosystems in response to changes in acid deposition and hydrology*, SC-DLO, Wageningen.

Latour, J.B., I.G. Staritsky, J.R.M. Alkemade and J. Wiertz (1997*). De Natuurplanner; Decision Support Systeem natuur en milieu; versie 1.1*, RIVM, Bilthoven.

Latour, J.B., R.J. van de Velde, H. van der Vloet, H. Veldhuizen, M. Ransijn and J. van de Waals (1998). *Methodiek voor het aangeven van de milieuhaalbaarheid van regionale gebiedsplannen: uitwerking voor het Groene Hart*, RIVM, Bilthoven.

MacEachren, A.M. (1994) *Some Truth with Maps: A primer on Symbolization & Design*, Association of American Geographers, Washington D.C.

Miller, G.A. (1956) The magical number seven plus or minus two: some limits on our capacity for processing information, in *Psychological Review*, Vol. **63**.

Monmonier, M.S. (1991) *How to lie with maps*, The University of Chicago Press, Chicago.

Muehrcke, P.C. and J.O. Muehrcke (1992) Map Use: *Reading.Analysis.Interpretation*, 3rd ed., JP Publications, Madison, Wisconsin.

Openshaw, S. (1991). Developing appropriate spatial analysis methods for GIS. In: D.J. Maguire, M.F. Goodchild and L.T. Steyaert (eds.), *Geographic Information Systems*, Longman Scientific & Technical, Harlow, pp. 389-402.

Pereira, J.M.C. and L. Duckstein (1993) A multiple criteria decision-making approach to GIS-based land suitability evaluation, *in International Journal for Geographical Information Systems*, Vol. 7, No. 5.

Tkach, R.J. and S.P. Simonovic (1997). A new approach to multi-criteria decision making in water resources. *Journal of Geographic Information and Decision Analysis*, Vol. 1, No. 1, pp. 25-43.

Tobler, W.R. (1973). Choropleth maps without class intervals? *Geographical Analysis*, Vol. 5, pp. 262-265

Tufte, E.R. (1985). *The visual display of quantitative information*, Graphic Press, Cheshire, Connecticut.

Tufte, E.R. (1990). *Envisioning information*, Graphics Press, Cheshire, Connecticut.

Unwin, D. (1981). *Introductory Spatial Analysis*, Methuen & Co. Ltd, London.

Upton, G.J.G. and B. Fingleton (1985). *Spatial Data Analysis by Example*, Wiley, New York.

17

Use Of Statistical Classification Techniques In Identifying Stock And Change In Broad Habitat Categories

Watkins, J.W., Howard, D.C. and Scott, W.A.

Data from the Countryside Surveys are being used to monitor changes in Biodiversity Action Plan Broad Habitats in Great Britain. The approach requires detailed data, collected independently of the Broad Habitat programme, to be reclassified into habitats and changes to be identified. In the surveys of 1984, 1990 and 1998 approximately 500 1 km squares were visited and mapped using a flexible coding system. The data held in GIS and databases allows statistical techniques to be used for measuring confidence in allocation and change in Broad Habitats between surveys.

Minkowski metrics are used to give a measure of confidence in the allocation of a land parcel to a habitat. The results reflect the variation due to the heterogeneity of different Broad Habitats and survey methodology. Although a single metric gives a measure of the variability in strength of affinity across all possible habitats, it is by comparing different orders of the Minkowski metric that confidence in the resulting allocation is quantified. It is possible to distinguish between an allocation due to a small number of large scores and one with a large number of smaller scores. Using second and third order metrics gives emphasis to single, large values as with a normal variance statistic. This enables confidence in allocations to be quantified.

Differences in the detailed strings of attribute codes describing a parcel in each survey may or may not indicate a change in Broad Habitat. Rather than use a simple comparison of the habitat allocations between different surveys to identify change, differences between scoring in all habitats quantified using Minkowski metrics provides a conservative estimate of change in habitat by rejecting change from *noisy* scoring.
The paper describes the calibration and application of the metrics and shows how the metric scores reflect the variability in parcels indicated by their attribute list and code weightings.

1. Introduction

Assigning cartographic elements to categories or classes is an essential prerequisite for many spatial analyses. Many different classification techniques can be used depending on the nature of the categories concerned and the aims of the classification procedure. Deterministic classification procedures use rule sets, keying procedures or expert judgement to allocate each element to a single category. In contrast, probabilistic classification procedures associate elements with several classes and quantify their strength of affinity to each class by some measure such as the statistical probability of membership. The value of multiple class associations is that it allows each element's characteristics to be represented in greater detail in a way that may better represent the reality of a heterogeneous world. However, such a scheme will complicate reporting the stock of elements in any particular class and will blur the definition of change from one class to another. This paper presents an approach to the reporting of change in elements with multiple class affinity.

Describing change in a system that uses deterministic class membership is simple, but highlights its weakness. Using a deterministic classification change is defined as an element shifting from one class to another. However, this definition masks differing degrees of change as measured by the internal composition of elements and exaggerates the importance of misclassification in borderline cases. To gain understanding of the nature of the changes occurring and to quantify levels of confidence in the measurement of change, data must be collected at a greater level of detail than the reporting classes. A class can then be characterised by the attributes of its constituent parts and a measurement of the strength of each element's affinity to the class made. It is the ability to measure changes in affinity strength within a class that allows change between classes to be defined more precisely.

Recording in greater detail than the reporting classes also provides the flexibility to be able to present results in other reporting schemes in the future. This is an important feature of the data set described in this paper, as it is used for long term monitoring of landscape features and has already been analysed in terms of several different classification schemes.

In the example described below, a set of data containing repeated observations of the same sample sites are summarised into a novel set of classes. An objective decision process is described that accepts or rejects a description of change dependent on the affinity of the element to different classes on different sampling occasions.

2.   The Broad Habitat Classes and Countryside Survey Data

In response to the Rio Convention, the UK government is committed to pursuing policies that increase biodiversity (HMSO, 1994). Part of this commitment is to audit and monitor biodiversity across Great Britain by the identification of a series of Broad Habitats that cover the whole of the land surface. These can then be managed through Biodiversity Action Plans (UK Biodiversity Steering Group, 1995). To assess and manage changes in biodiversity, a knowledge of the history of change and current trends is invaluable. The procedure involves the reworking of data collected before the Broad Habitats were conceived and presenting them in the new classification. Seventeen terrestrial Broad Habitats have been defined in GB with a further 10 littoral and marine Habitats.

The Countryside Surveys (Barr *et al.*, 1993) are a series of national surveys of GB made during 1978, 1984, 1990 and 1998. Each survey consisted of visiting, mapping and sampling a set of 1 km squares selected from an environmental stratification of the whole of GB (Bunce *et al.*, 1996). Field surveyors mapped all land cover of each square into five potentially overlapping themes; agriculture & semi-natural vegetation, forestry, physiography, buildings & communications and boundary features. The features of each distinct parcel of land were recorded using a variable number of standard attribute codes. The resulting maps of each square were digitised as polygon coverages in a GIS and linked to a database holding the attribute codes for each polygon (Howard & Barr, 1991). Within each theme there are detailed records including agricultural use and management, feature condition and age and species abundance.

To allocate Countryside Survey land parcel polygons to Broad Habitat summary classes for reporting purposes, a matrix was produced giving each field attribute code a score describing its potential occurrence in each of the classes. For example, the code for oak would score highly in broadleaf and mixed woodland but would have no score in conifer wood or calcareous grassland. All polygons can be classified by summing the scores of its attribute codes within each summary class. The polygons are then allocated to the class with the highest score. If several classes tie, the polygon can be left as a mosaic of types or allocated to a particular class using a rule base. The basic procedure for this allocation is illustrated in Table 1.

In the example presented in Table 1, two sets of attribute codes describing a parcel in two different years are presented along with their weightings for the 22 Broad Habitats. Summing the weightings for each Broad Habitat and identifying the largest produces an allocation of Neutral Grass for the first year and Acid Grass for the second.

3. Measuring class affinity and confidence in allocation

In the procedure outlined above, a parcel of land will have attributes that score in several different Broad Habitat classes. This profile of Broad Habitat scores for each land cover element shows whether it has associations with only one class, such as Conifer Woodland, or of several classes, such as the grassland types. Furthermore the magnitude of any one score can be taken as an indicator of the strength of affinity to a class and collectively they describe the diversity present in the polygon. Hence, an element may be said to have a strong or weak affinity to one or several Broad Habitats. Weak affinity may indicate that the characteristics of the land cover element are poorly described by the Broad Habitat classification or may indicate a paucity of information from the survey. The overall polarity of a profile may be used as a measure of confidence in an element's final class allocation. Collecting more information to strengthen a profile may not give a clear allocation but may broaden the number of classes with which an element shows affinity. Confidence in the classification is therefore a function of the breadth of an element's affinities as well as their strengths.

In order to quantify confidence in the allocation of an element to any one class, the magnitude of all the element's affinities needs to be examined. The greater the dissimilarity in the profile of scores, the greater the confidence in allocating an element to one particular Broad Habitat class.

In the current example of classification of land cover elements into Broad Habitat classes, the highest scoring Broad Habitat in the element's profile determines its allocation. The level of confidence will not alter the allocation of an element to a Broad Habitat. However, a measure of confidence does affect allocation when determining Broad Habitat change.

ATTRIBUTE CODE NUMBER	ATTRIBUTE CODE DESCRIPTIONS	BROADLEAF, MIXED AND YEW	CONIFER	LINEAR	ARABLE	IMPROVED GRASS	NEUTRAL GRASS	CALCARIOUS GRASS	ACID GRASS	BRACKEN	SHRUB HEATH	FEN	BOG	OPEN WATER	RIVER	MONTANE	INLAND ROCK	URBAN	SUPRALITTORAL ROCK	SUPRALITTORAL SEDIMENT	LITTORAL ROCK	LITTORAL SEDIMENT	SEA
						Scoring for parcel in 1990 survey																	
101	Lowland agricultural grass					5	4																
154	*Agrostis capillaris*						1		7														
176	50-75%																						
152	*Cynosurus cristatus*					1	7																
175	25-50%																						
186	Dairy					1																	
194	Hay					2	1																
	Broad Habitat Scores - 1990	0	0	0	0	9	13	0	7	0	0	0	0	0	0	0	0	0	0	0	0	0	0
						Scoring for parcel in 1998 survey																	
101	Lowland agricultural grass					5	4																
154	*Agrostis capillaris*						1		7														
176	50-75%																						
197	No apparent use																						
	Broad Habitat Scores - 1998	0	0	0	0	5	5	0	7	0	0	0	0	0	0	0	0	0	0	0	0	0	0

Table 1 *Allocation of a single parcel to Broad Habitats using attribute codes*

4. Comparing class affinities and defining class change

Where elements have been assigned to a set of classes, the measurement of change over time needs to be defined relative to both the individual elements and the overall classification. Where the characteristics of elements have been defined independently of the classification, it is possible to observe changes in the composition of an element that may not affect its class assignment. Measurement of these individual changes can be used to define what is a significant change in class affinity and also give confidence in assigning change to any particular element.

As stated above the confidence in allocation of an element to a class is a function of the strength and breadth of its affinities, as measured in this case by the elements Broad Habitat scores. Comparison of an element's scores at two points in time can be achieved in a variety of ways. One common method is to use a generalised distance measure such as the Minkowski metric -

$$\left\{ \sum_{k=1}^{p} \left| x_{ik} - x_{jk} \right|^{\lambda} \right\}^{1/\lambda} \quad \textbf{Equation 1}$$

where x_{ik} is the score of the kth class at time one and x_{jk} is the score of the kth class at time two. The parameter λ can be chosen to suit the requirements of particular applications. $\lambda = 1$ gives the *city block* metric, while $\lambda = 2$ gives the Euclidean metric used in *residual mean square* calculations (Krzanowski, 1988). Increasing λ emphasises the more dissimilar scores where the scores are now the difference in affinity between the two surveys.

Using the Minkowski metric a larger value indicates greater confidence in change in the elements class affinities. If a high value is combined with a change in the highest scoring class between surveys the element can be defined as having changed class allocation. However, if the metric gives a low value, the element will not be assigned change regardless of a change in highest scoring class. This is because the metric value indicates that there is insufficient evidence that a real change has taken place in the elements underlying characteristics. This *override* of the class allocation by the metric value is most likely to take place in borderline cases where confidence in individual class allocations is low at the first, second or both time points. The definition of borderline in this case will be allied to the definition of a sufficiently high metric score to indicate change. By limiting the change to where it is clear-cut, the process is conservative.

5. Calibrating the change metric

To define the metric value at which change will be assigned the underlying structure of the code scoring system must be examined. In the current example each attribute code used to record the characteristics of land cover elements has been given a weight for each Broad Habitat. Codes receive different weights depending on how indicative they are of a particular Broad Habitat. Codes fall into three categories of decreasing importance:
i)   codes describing landscape type (e.g. agricultural grass),
ii) species or other specific indicator codes and
iii) more ubiquitous species and weaker indicator elements.

Generally, weights of 10, 5 and 1 were given to these codes. Empirically this was found to produce sufficient separation in habitat scores to distinguish those elements whose primary or indicator species codes have changed from those where only minor species or feature codes have changed. At the same time the weights are sufficiently close that the evidence from codes in the second and third groups can accumulate to the extent that they are considered indicative of a significant change.

In order to test whether different types of change in element score profiles could be detected a test data set was generated with one group of elements showing change in minor codes only, one showing change in minor codes and indicator species, and one showing change in all code types. Minkowski metrics of different degree were tested to identify the value of λ best suited to discriminate the different types of change.

Using the simulated data it was found that the best discriminator was not an individual form of the metric but the difference in two consecutive forms -

$$M = \left\{ \frac{1}{P} \sum_{k=1}^{P} \left| x_{ik} - x_{jk} \right|^{\lambda+1} \right\}^{1/(\lambda+1)} - \left\{ \frac{1}{P} \sum_{k=1}^{P} \left| x_{ik} - x_{jk} \right|^{\lambda} \right\}^{1/\lambda} \quad \textbf{Equation 2}$$

where the best discrimination was found at $\lambda = 2$. The denominators, P, were introduced to facilitate comparison of different metric powers.

The underlying weighting of the attribute codes gives rise to a clear division between elements with changes in minor codes and those with changes in primary and indicator species. The division occurred at a metric score of 0.5 (see Figure 1)

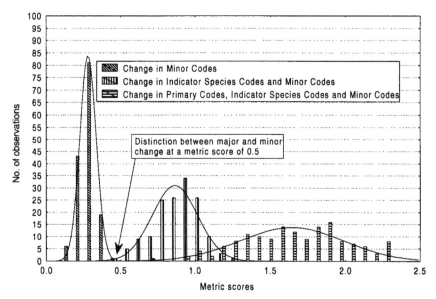

Figure 1. *Modelled data sets displaying different change characteristics to enable calibration of the metric*

The metric developed from the simulated data was then used to examine score changes in the actual field data examining approximately 40,000 parcels from the 500 1 km squares from across GB. The same pattern of division was observed across all areas of the data sets and in all Broad Habitats (see Figure 2). Hence, the metric given in equation 2 and the indicated division point was adopted for measuring confidence in change for all land cover elements.

Table 2 shows the frequency with which changes between different Broad Habitat classes were rejected. There were just over 1300 cases where the Minkowski metrics lead to an override of the change (3%). It is clear that some classes are more prone to relatively unimportant variation than others. Almost none of the changes into or out of *Calcareous Grass*, for example, are rejected whereas a relatively high proportion of the changes into and out of *Improved Grass* are rejected particularly those to and from

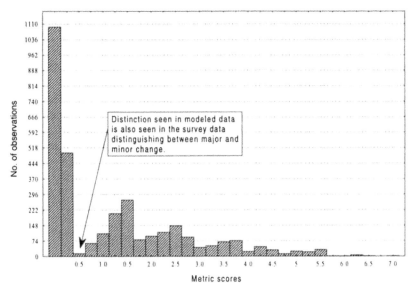

Figure 2. *Metric scores from survey data reflecting change in features between CS1990 and CS2000*

Broadleaf, Mixed and Yew and *Neutral Grass*. The reasons for this are complex, including the definition of the different classes (*Calcareous Grass* has strong indicators), the size of the classes in both years (*Calcareous Grass* is smaller than *Improved Grass*) and the potential for change (*Calcareous Grass* is a more stable, semi-natural class while *Improved Grass* is under close management).

6. Discussion

There are benefits to using a standard statistical method to determine the significance of change, but it also has specific consequences. Three descriptions of a parcel are being made, first its Broad Habitat at time one, then its Broad Habitat at time two and finally the change from one Broad Habitat to another. For estimates of change derived solely from a simple deterministic allocation, the third description can be derived from the other two, but in this case that is not necessarily so. The best estimate for the Broad Habitat at either time point is still that produced by looking at each individual years data, but the additional detailed element data may suggest that no real change has occurred in a parcel classified into different Broad Habitats in the two years. The problem is largely presentational; the Minkowski derived change statistics are practically useful, as they are conservative estimates of where clear change has occurred.

Ancillary statistics, such as the difference between the best individual time estimates and the correspondence matrix Table 2 should offer an aid to interpretation. For example, *Calcareous Grass* has a far smaller number of changes rejected than *Improved*

Grass. This indicates the quality of change, contrasting dramatic shifts with more subtle evolution and suggests where the edges of the Broad Habitat definitions are less clear.

The clarity or confidence in allocation is available to the GIS through each parcel's Broad Habitat scores. The profile of these scores reflects the underlying heterogeneity indicated by the attribute codes, which is hidden by allocating the parcel to a single Broad Habitat category. This underlying information could be used in graphical presentation of confidence calculated for each parcel using the highest score vs. the total of all scores (c.f. maximum likelihood). Alternatively, a representation of heterogeneity could be presented by allocating areas within each parcel according to its scoring in different Broad Habitats (Aspinall and Pearson, 1995). Information from the scoring profile of each parcel is especially important in scenario testing and permutation methods for predicting change. Use of this information in providing confidence measures for predicted change as well as observed change is the subject of ongoing research.

The analysis of change, especially of repeated measures, needs more careful interpretation than the production of single measures. Where decisions are likely to be made using the results, they need to be conservative and robust. Decisions taken producing the statistics have to be objective and transparent and the final presentation should be qualified by measures of confidence. In this case, a procedure, the Minkowski metric, has been employed that meets those criteria.

Acknowledgements

The Department of Environment, Transport and the Regions co-funded of the Countryside Surveys. We thank Colin Barr, the Countryside Survey field survey project leader for valuable direction and discussion and Morna Gillespie for managing and organising the GIS.

References

Aspinall, R. J. and Pearson, D. M. (1995) Describing and managing uncertainty of categorical maps in GIS. In *Innovations in GIS 2* ed. P. Fisher p 71-83 Taylor and Francis, London.

Barr, C.J., Bunce, R.G.H., Clarke, R.T., Fuller, R.M., Furse, M.T., Gillespie, M.K., Groom, G.B., Hallam, C.J., Hornung, M., Howard, D.C. & Ness, M.J. (1993) *Countryside Survey 1990 main report* Department of the Environment, London.

Bunce, R.G.H., Barr, C.J., Clarke, R.T., Howard, D.C. & Lane, A.M.J. (1996) The ITE Merlewood Land Classification of Great Britain *Journal of Biogeography* **23,** 625-634.

UK Biodiversity Steering Group,. (1995) *Biodiversity: the UK Steering Group report* HMSO, London.

HMSO, (1994) Biodiversity: the UK Action Plan. HMSO, London

Howard, D.C. & Barr, C.J. (1991) Sampling the countryside of Great Britain: GIS for the detection and prediction of rural change. In *Applications in a changing world*, vol. 153. *FRDA report*, pp. 171-176. Forestry Canada., Ottawa.

Krzanowski, W.J. (1988) *Principles of Multivariate Analysis* Oxford University Press, Oxford.

Broad Habitat class 2000 (rows) × _Broad Habitat class 1990_ (columns)

Broad Habitat class 2000 ↓ \ 1990 →	BROADLEAF, MIXED & YEW	CONIFER	LINEAR	ARABLE	IMPROVED GRASS	NEUTRAL GRASS	CALCARIOUS GRASS	ACID GRASS	BRACKEN	SHRUB HEATH	FEN	BOG	OPEN WATER	RIVER	INLAND ROCK	URBAN	SUPRALITTORAL ROCK	SUPRALITTORAL SEDIMENT	LITTORAL SEDIMENT	SEA
SEA													0.2							
LITTORAL SEDIMENT				0.1																
SUPRALITTORAL SEDIMENT	0.2															0.2				
SUPRALITTORAL ROCK	0.2										0.1	0.4				0.1	0.2			
URBAN	1.4	0.1	1.1	0.2	1.9	1.5		0.3			0.1				0.1					
INLAND ROCK	0.1	0.1						0.3					0.1							
RIVER	0.1												0.2							
OPEN WATER	0.5							0.8			0.2	0.5				0.3				
BOG								0.1		0.4										
FEN					0.3	0.2				0.1						0.2				
SHRUB HEATH																				
BRACKEN					0.2											0.1				
ACID GRASS	0.2				1.1	0.5										0.2				
CALCARIOUS GRASS																				
NEUTRAL GRASS	4.4	2.3	0.9	0.8	6.5			0.8	0.3		0.2	0.1		0.2		1.0			0.4	
IMPROVED GRASS	5.8	0.2	1.0	1.8		3.2		3.7		0.2	4.4		0.1	0.1		6.3			0.4	
ARABLE			0.1		1.1	0.6		0.1			0.1					0.5				
LINEAR	0.7			0.4	0.5								0.1							
CONIFER	2.3				0.1	0.2		0.6	0.2	2.2	0.5	1.4				0.2				
BROADLEAF, MIXED & YEW	2.3	0.6		1.7	6.5	5.1	0.2	0.9	0.5	0.1	0.5	0.3	1.0	0.2	0.1	2.3	0.2			

Table 2 Percentage of elements showing a change in Broad Habitat allocation whose change was rejected using the Minkowski metric

Index

9 780415 253628